Orbital Angular Momentum States of Light (Second Edition)

Propagation through atmospheric turbulence

Online at: https://doi.org/10.1088/978-0-7503-5959-7

IOP Series in Advances in Optics, Photonics and Optoelectronics

SERIES EDITOR

Professor Rajpal S Sirohi Consultant Scientist

About the Editor

Rajpal S Sirohi is currently working as a faculty member in the Department of Physics, Alabama A&M University, Huntsville, AL, USA. Prior to this, he was a consultant scientist at the Indian Institute of Science, Bangalore, and before that he was Chair Professor in the Department of Physics, Tezpur University, Assam. During 2000–2011, he was an academic administrator, being vice-chancellor to a couple of universities and the director of the Indian Institute of Technology, Delhi. He is the recipient of many international and national awards and the author of more than 400 papers. Dr Sirohi is involved with research concerning optical metrology, optical instrumentation, holography, and the speckle phenomena.

About the series

Optics, photonics, and optoelectronics are enabling technologies in many branches of science, engineering, medicine, and agriculture. These technologies have reshaped our outlook and our ways of interacting with each other, and have brought people closer together. They help us to understand many phenomena better and provide deeper insight into the functioning of nature. Further, these technologies themselves are evolving at a rapid rate. Their applications encompass very large spatial scales, from nanometers to the astronomical scale, and a very large temporal range, from picoseconds to billions of years. This series on advances in optics, photonics, and optoelectronics aims to cover topics that are of interest to both academia and industry. Some of the topics to be covered by the books in this series include biophotonics and medical imaging, devices, electromagnetics, fiber optics, information storage, instrumentation, light sources, charge-coupled devices (CCDs) and complementary metal oxide semiconductor (CMOS) imagers, metamaterials, optical metrology, optical networks, photovoltaics, free-form optics and its evaluation, singular optics, cryptography, and sensors.

About IOP ebooks

The authors are encouraged to take advantage of the features made possible by electronic publication to enhance the reader experience through the use of color, animation, and video and by incorporating supplementary files in their work.

A full list of titles published in this series can be found here: https://iopscience.iop.org/bookListInfo/series-on-advances-in-optics-photonics-and-optoelectronics.

Orbital Angular Momentum States of Light (Second Edition)

Propagation through atmospheric turbulence

Kedar Khare
Optics and Photonics Centre, Indian Institute of Technology Delhi, New Delhi, India

Priyanka Lochab
Department of Applied Sciences and Humanities, Indira Gandhi Delhi Technical University for Women, New Delhi, India

Paramasivam Senthilkumaran
Optics and Photonics Centre, Indian Institute of Technology Delhi, New Delhi, India

IOP Publishing, Bristol, UK

ISBN 978-0-7503-5959-7 (ebook)
ISBN 978-0-7503-5957-3 (print)
ISBN 978-0-7503-5960-3 (myPrint)
ISBN 978-0-7503-5958-0 (mobi)

DOI 10.1088/978-0-7503-5959-7

Version: 20241201

IOP ebooks

British Library Cataloguing-in-Publication Data: A catalogue record for this book is available from the British Library.

Published by IOP Publishing, wholly owned by The Institute of Physics, London

IOP Publishing, No.2 The Distillery, Glassfields, Avon Street, Bristol, BS2 0GR, UK

US Office: IOP Publishing, Inc., 190 North Independence Mall West, Suite 601, Philadelphia, PA 19106, USA

Contents

Preface

Preface to the second edition

Writing the second edition of our book *Orbital Angular Momentum States of Light (Second Edition): Propagation through atmospheric turbulence* has been a rewarding experience. We are grateful for the positive global reception of the first edition from readers, especially from research students. Numerous readers reached out to us with thoughtful queries and also expressed appreciation for the inclusion of annotated turbulence propagation code in the book. Writing of the first edition of the book was initiated while the second author (PL) was a senior PhD student and our research work in the topical area of beam propagation in atmospheric turbulence had been initiated essentially for her thesis. Over the last few years, we have engaged ourselves further with experiments on engineered laser beam propagation through turbulence for real world applications of high value. This has provided us with a deeper and more refined understanding of the subject, which we have thoughtfully integrated into the Second Edition.

The most significant addition to the Second Edition is chapter 10, which opens up a new line of investigation. During our investigations on the speckle phenomenon with engineered beams, a surprise emerged that the speckle intensity patterns show distinct textural properties depending on the state of the illuminating vector beam. While speckle literature has evolved following the invention of laser, most statistical studies have focused on point statistics of intensity or phase. The discovery that speckles can exhibit distinct textural properties introduces new insights with promising potential for future applications, such as real-time atmospheric characterization. We believe that these new findings certainly deserved attention, which is why we have dedicated a full-length chapter to explore them in detail. Additionally, we have filled the gaps which existed in the first edition by including discussion on the coherent mode description of partially coherent light and its connection with propagation of engineered beams. This discussion offers some new important pointers for designing higher order robust beams which we cover in chapter 9. This required us to introduce some additional mathematical ideas in the early part of the book in chapter 2. Furthermore, we have included a brief discussion comparing the on-axis intensity statistics of scalar and vector beams, which holds significant implications for applications such as free-space communication.

KK would like to thank IIT Delhi administration for the supportive academic environment where a book writing exercise could be taken up. PL started a new faculty position at Indira Gandhi Delhi Technical University For Women, New Delhi. Completing the Second Edition of the book while simultaneously handling the challenges of a new faculty position has given her a very good perspective as she hopes to initiate her independent research along with continuation of collaborative efforts with the IIT Delhi group. We deeply appreciate the encouragement from Professor Rajpal S Sirohi, the editor of this IoP series, for writing this book. We are also grateful for the professional support and patience shown by IoP editorial office, particularly Ms Ashley Gasque and Ms Bethany Hext, while the manuscript for the Second Edition was being prepared. We additionally thank IoP for making an electronic copy of the First Edition of this book available to the participants of the 6th International Conference on Optical Angular Momentum (ICOAM) held in Finland in 2022, which gave it a good visibility in the community. The book was entirely written using the collaborative LaTeX environment provided by Overleaf to which our institute subscribes. Finally, we are deeply indebted to our respective families for their continuous support and patience during this book writing effort.

Kedar Khare
Priyanka Lochab
Paramasivam Senthilkumaran
New Delhi
August 2024

Preface to the first edition

This book has origins in the present authors' work over years on a range of topical areas including complex signal representation in Optics, propagation characteristics of phase and polarization singularities, understanding the near core structure of optical vortices, quantitative phase imaging and light propagation in random media. Over the last five years, one of the authors (PL) worked under the supervision of the other two (KK and PS) for her PhD thesis at Indian Institute of Technology Delhi, India and our lively interactions allowed us to have a re-look at these interesting topics through a beginning researcher's viewpoint. Laser beam propagation in turbulence is a mature area with literature spanning over more than 70 years. A large number of publications and some good books have been devoted to this topic that is of immense interest to a number of practical applications like defense systems and free space classical and quantum communication.

PL's recently completed thesis work explored the question of whether one can utilize the degrees of freedom such as amplitude, phase, polarization, and orbital angular momentum (OAM) of light for designing laser beams that can inherently maintain robust intensity profiles on long range propagation through turbulence. While investigating this question we gained several valuable insights. At the same time we constantly felt the need for a single book/monograph that explained the key details regarding modeling of optical beam propagation through atmospheric turbulence and at the same time delved into more exotic topics like OAM states of light and polarization singularities that are essential for understanding generation and applications of structured light beams. So when it was time for PL to put together her PhD thesis document, we took up the exercise of writing this book in parallel. We believe that the material in this book will be helpful to optics students, researchers and engineers working on long range laser propagation systems in two important ways. It will give them a historical perspective on our current understanding of light propagation through turbulence, and at the same time provide computer modeling tools for making realistic assessment of arbitrary beam profiles after long range propagation. The pieces of computer code explained in the appendix A of this book can be run in open source environment like GNU Octave and may also be readily modified to suit individual systems and turbulence conditions. So in some sense this book is also an attempt to de-mystify the topic of beam propagation in turbulence and allow beginning researchers to get a quick start into the real problems they are interested to address. We welcome readers to try out the codes and contribute toward improving them further. KK would like to thank IIT Delhi administration for allowing a sabbatical year that allowed him to focus on writing this book without too many administrative distractions. PL would like to acknowledge the fellowship support and the open research environment at IIT Delhi for pursuing her recently concluded PhD thesis work. We deeply appreciate the encouragement from Prof. Rajpal S Sirohi, the editor of this IoP series, for writing this book. We are also grateful for the professional support and patience shown by IoP editorial office, particularly Ms Ashley Gasque, while the manuscript for this book was being prepared. The book was entirely written using the collaborative LaTeX environment provided by Overleaf to which our institute subscribes. Finally we are deeply indebted to our respective families for their continuous support and patience during this book writing effort.

Kedar Khare
Priyanka Lochab
Paramasivam Senthilkumaran
New Delhi
September 2020

Author biographies

Kedar Khare

Dr Kedar Khare is a joint Professor in the Optics and Photonics Centre and Department of Physics, Indian Institute of Technology Delhi, New Delhi, India. He currently holds the Abdul Kalam National Innovation Fellowship constituted by Indian National Academy of Engineering (INAE). Prior to joining the faculty of IIT Delhi in December 2011, he spent several years working as a Lead Scientist in the Imaging Technologies division at General Electric Global Research, Niskayuna, USA. Professor Khare's research interests span a wide ranging problems in computational optical imaging. In recent years he has been actively engaged in pursuing research work on quantitative phase imaging, Fourier phase retrieval, computational microscopy, diagnostic imaging, cryo-EM imaging and structured light propagation in turbulence. Professor Khare also serves as Editor-in-Chief for *Journal of Modern Optics* published by Taylor and Francis, UK. Professor Khare received his Integrated MSc in Physics from Indian Institute of Technology Kharagpur with highest honours and his PhD in optics from The Intitute of Optics, University of Rochester, USA.

Priyanka Lochab

Dr Priyanka Lochab is an Assistant Professor in the Department of Applied Sciences and Humanities at Indira Gandhi Delhi Technical University for Women (IGDTUW). She started her research journey in 2013 at the premier Indian Institute of Technology (IIT) Delhi, where she completed her PhD and later served as a Research Associate in the Department of Physics. Her research interests include singular optics, beam propagation through random media, and computational imaging. Dr Lochab has been recognized for her contributions to the field with awards such as the Best Thesis Award by the Optical Society of India (OSI) and the Distinction in Research by IIT Delhi. As a young and emerging scientist, she continues to explore new frontiers in optical science, striving to make impactful contributions to the field.

Paramasivam Senthilkumaran

Dr Paramasivam Senthilkumaran is working as a full Professor with a joint appointment at the Optics and Photonics Centre ands Department of Physics, Indian Institute of Technology Delhi (IIT Delhi). He currently holds the Ajoy Ghatak Chair Professorship. He joined IIT Delhi as an Assistant Professor of Physics in 2002 and became Associate Professor in 2008 and Professor in 2012. He worked as Assistant Professor between 1996 and 2001 and as a lecturer between 1995 and 1996 in IIT Guwahati (IITG). Prior to joining IITG, he had been a senior project officer since September 1993 at IIT Madras from where he received his PhD in 1995. He is a recipient of the Young Scientist Award from Indian National Science Academy (INSA), New Delhi and Alexander von Humboldt fellowship, Germany in 1997 and 2001, respectively. He was in University of Strathclyde, Glasgow, United Kingdom on a Royal Society-London and INSA exchange fellowship in 1999 and in Friedrich Schiller University of Jena, Germany during 2001–2002 on a Humboldt fellowship.

He has been teaching undergraduate and postgraduate courses on basic physics, electromagnetic theory, optics and lasers, Fourier optics, holography and its applications, optical metrology and optical instrumentation. His research interests are optical beam shaping, optical phase singularities, Berry and Pancharatnam topological phases, fiber optics, holography, non-destructive testing techniques, shear interferometry, Talbot interferometry, speckle metrology and non-linear optics. He has authored/coauthored more than one hundred research publications. He has authored one book entitled *Singularities in Physics and Engineering* published by IOP Publishing, Bristol, UK in 2018. He has guest edited a special issue on singular optics for *International Journal of Optics* in 2012 along with Professor Shunichi Sato and Professor Jan Masajada. Professor Paramasivam Senthilkumaran has also been an editorial board member of *International Journal of Optics* since 2013 and associate editor of *Optical Engineering* since 2020.

Chapter 1

Introduction

The orbital angular momentum (OAM) states of light have gained importance over the last few decades as a promising choice of modes for free-space communication as well as in defence systems. These applications inherently require long-range propagation of light beams through a fluctuating atmosphere. Various local factors such as temperature, pressure, humidity, wind velocity, etc, lead to space- and time-varying refractive index profiles that are nearly impossible to control over kilometers of distances. Even small changes in local refractive index ($\approx 10^{-5}$–10^{-6}) can have significant undesirable effects on a carefully designed laser beam after long-range propagation through the atmosphere. Therefore, understanding and modeling of propagation of light beams with arbitrary amplitude, phase and polarization profiles through a given set of atmospheric turbulence conditions is of utmost importance for practical system designers.

The literature on laser beam propagation through atmospheric turbulence spans over more than 70 years. In recent decades, several exciting developments have occurred in this area, including advanced active beam correction methods employing adaptive optics, the potential use of structured light beams for robust beam engineering, and the possibility of using entangled states of light for long-range communication with satellites, to name a few. The use of OAM states of light has been a common feature in these developments, thus creating renewed interest in the detailed study of light propagation through atmospheric turbulence. For both a beginning researcher at the graduate student level and a seasoned optical engineer, sifting through the vast literature in this topical area is a daunting task. Most prior results in this area have been reported for Gaussian beams although the literature on structured light propagation in the atmosphere is growing. Further, testing of newer beam engineering ideas in this topical area is not easy as there is a lack of standalone, easily accessible and inexpensive software tools in the public domain that allow rigorous modeling of propagation of beams with arbitrary field profiles through turbulence. The aim of this book and the accompanying software code

doi:10.1088/978-0-7503-5959-7ch1

pieces is to address these difficulties and make the subject accessible to a larger scientific community.

The book is organized in two main parts. In the first part we provide discussion on the basic mathematical tools used throughout the book. Chapter 2 discusses topics such as Fourier and Hilbert transforms, basics of random processes and inverse probability law for generating realizations of random processes with given spectral properties. An important result which generalizes the notion of Hilbert transform to two dimensions is also presented in chapter 2 which forms the basis of the novel robust beam engineering principle described later. One important aspect of this book is the extensive use of the angular spectrum approach for field propagation calculations, which we believe is superior to the more commonly used Fresnel propagation method. The angular spectrum method is described in detail in chapter 3. As an illustration of the angular spectrum method, we discuss the propagation characteristics of a phase vortex, with particular attention to its near-core structure in chapter 4. Propagation characteristics of the OAM states as well as vector beams containing polarization singularities through free space are described in chapters 5 and 6, respectively. Polarization singularities are often presented in optics literature as exotic objects. However we present this topic in a more accessible manner and develop them simply as superposition of OAM states in an orthogonal polarization basis. This point of view is useful when discussing diversity mechanisms associated with propagation of structured light in atmospheric turbulence.

The second part of the book deals with the description of optical properties of a fluctuating atmosphere. The historical developments on modeling of wave propagation in turbulence is described in chapter 7 with reference to the Born and Rytov approximations and the extended Huygens–Fresnel principle. The numerical methods for simulating beam propagation through the atmosphere by means of a multiple-phase screen model is described in chapter 8. In these two chapters, we provide important details on various refractive index spectrum profiles (such as Kolmogorov, von Karman), sampling criteria related to atmospheric fluctuation scales, considerations for choosing the number of random phase screens depending on the turbulence strength, and the inclusion of sub-harmonic components in phase screens. The discussion is presented in a way that readers can readily implement these aspects in their work. Finally, in chapter 9 we describe simulations and experimental results on propagation of optical beams through long-range turbulence. Both scalar and vector beams are considered, with particular attention paid to the propagation of polarization singularities through turbulence. The work of the present authors on engineering of polarization singular beams that can maintain robust intensity profile on propagation through turbulence is discussed in this context by using speckle diversity as a guiding principle. Future directions and experiments that can effectively utilize the material presented in this book are also discussed. We describe the rationale behind traditionally used beam quality parameters such as scintillation index and their limitations in studying the characteristics of structured beams and further introduce instantaneous beam signal-to-noise ration (SNR) as a new beam quality measure that may be more suited to understanding propagation of structured light beams through turbulence. In a similar vein, in chapter 10, we describe novel ideas associated with texture of speckle observed when structured light is propagated through a random medium. Speckle statistics is traditionally treated as a single observation point statistics. When speckle is produced by structured light, we additionally describe interesting results which show that the two-dimensional patterns characteristic of the illumination are seen in the resultant speckle. These texture patterns can find applications in designing a novel communication channel mechanism through random media or for characterizing the properties of the random media themselves. Several pieces of annotated computer code with brief explanations are provided in appendix A. The reader can quickly incorporate or modify the computer codes as suited to their specific system modeling needs depending on the turbulence conditions of interest.

The coherent mode representation for partially coherent light is introduced early on and then used later to motivate the design of turbulence resilient beams. The major material in the book is, however, mostly presented in terms of scalar fields (or their combinations) because the corresponding interpretations based on space-frequency correlation functions are sometimes harder to interpret for beginning researchers or practicing optical engineers.

Also, not much space is devoted to communication using OAM modes as this is not the main thrust of this book. The change of scintillation statistics due to scalar and structured light beams is, however, highlighted, as it has important implications for improving the error rate performance of free-space optical communication systems. We believe that the researchers involved in work on free-space (classical as well as quantum) optical communication will find the details of turbulence propagation modeling useful for evaluating their new ideas. The topics covered in this book along with extensive survey of literature and computer codes can be covered as special topics in a graduate level course or can simply be treated as a reference material. The simulation codes are consciously written in a way that they can be used with proprietary software like MATLAB as well as with its open source clone GNU Octave. The codes do not use any software specific features and can be quickly adapted to model a number of practical long-range propagation scenarios without difficulty. We believe that with these features, the book will be useful to a number of researchers and optical engineers.

IOP Publishing

Orbital Angular Momentum States of Light (Second Edition)
Propagation through atmospheric turbulence
Kedar Khare, Priyanka Lochab and Paramasivam Senthilkumaran

Chapter 2

Mathematical preliminaries

In this chapter we describe some basic mathematical tools that will be used throughout the book. The material covered here will also help to set the basic notations. The topics discussed here include results on the Fourier transform and random processes that are covered in excellent textbooks. Some specialized results such as quadrature representation for two-dimensional signals will be discussed in detail as they will be useful in later chapters of the book.

2.1 Fourier transform basics

Fourier transform is an important tool that we will be using often in this book in the context of propagation of wave fields. For a finite energy function $g(x, y)$, we will use the following definition for the Fourier transform pair:

$$G(f_x, f_y) = \mathcal{F}[g(x, y)] = \int_{-\infty}^{\infty}\int_{-\infty}^{\infty} dx\, dy\; g(x, y)\exp[-i2\pi(f_x x + f_y y)], \quad (2.1)$$

and

$$g(x, y) = \mathcal{F}^{-1}[G(f_x, f_y)] = \int_{-\infty}^{\infty}\int_{-\infty}^{\infty} df_x\, df_y\;\; G(f_x, f_y)\exp[i2\pi(f_x x + f_y y)]. \quad (2.2)$$

The definition in equations (2.1) and (2.2) suggests that the function $g(x, y)$ is expressible as a linear combination of complex exponentials, with the Fourier transform $G(f_x, f_y)$ denoting the weight corresponding to the term $\exp[i2\pi(f_x x + f_y y)]$. Some of the well-known properties of the Fourier transform that we will find useful are:

1. **Linearity**: The Fourier transform is a linear operator. For constants c_1 and c_2,

$$\mathcal{F}[c_1\, g_1(x, y) + c_2\, g_2(x, y)] = c_1\, G_1(f_x, f_y) + c_2\, G_2(f_x, f_y). \quad (2.3)$$

2. **Scaling**: The Fourier transform of the scaled function $g(ax, by)$ is given by:

$$\mathcal{F}[g(ax, by)] = \frac{1}{|a||b|} G\left(\frac{f_x}{a}, \frac{f_y}{b}\right). \tag{2.4}$$

The stretching of a function in (x, y) space thus causes compression of its Fourier transform and vice versa.

3. **Shift**: The Fourier transform of a shifted function $g(x - a, y - b)$ is given by:

$$\mathcal{F}[g(x - a, y - b)] = G(f_x, f_y)\exp[-i2\pi(f_x a + f_y b)]. \tag{2.5}$$

Translation of a function is thus equivalent to multiplication of its Fourier transform by a phase ramp.

4. **Energy theorem**: The energy of a function in (x, y) space is conserved in the two-dimensional Fourier space. Thus,

$$\int_{-\infty}^{\infty}\int_{-\infty}^{\infty} dx\, dy\, |g(x, y)|^2 = \int_{-\infty}^{\infty}\int_{-\infty}^{\infty} df_x\, df_y\, |G(f_x, f_y)|^2. \tag{2.6}$$

5. **Convolution**: The convolution operation of two functions in (x, y) space is equivalent to the product of their Fourier transforms We will often use this property later in the book when describing free-space diffraction of a beam.

$$\mathcal{F}\left[\int_{-\infty}^{\infty}\int_{-\infty}^{\infty} dx'\, dy'\, g(x', y')h(x - x', y - y')\right] = G(f_x, f_y)H(f_x, f_y). \tag{2.7}$$

These properties have been stated here without proofs. The proofs can be found elsewhere in excellent resources on Fourier transform theory [1, 2]. For completeness we will introduce the notation for delta impulse which may be represented as:

$$\delta(x, y) = \int_{-\infty}^{\infty}\int_{-\infty}^{\infty} df_x\, df_y\, \exp[i2\pi(f_x x + f_y y)]. \tag{2.8}$$

The delta impulse is a generalized function or distribution and the sampling property associated with it is given by:

$$\int_{-\infty}^{\infty}\int_{-\infty}^{\infty} dxdy\, \delta(x - x_0, y - y_0)g(x, y) = g(x_0, y_0). \tag{2.9}$$

Multiplication of $g(x, y)$ with a delta impulse located at $x = x_0$, $y = y_0$ followed by integration over the whole (x, y) plane thus provides the value of the sample $g(x_0, y_0)$ of the function $g(x, y)$.

A signal $g(x, y)$ is called band-limited if its Fourier transform $G(f_x, f_y)$ is zero outside some spatial frequency interval $2\Omega_x \times 2\Omega_y$. Practically most signals will have negligible energy outside a certain frequency band. Band-limited signals admit a sampling expansion of the form:

$$g(x, y) = \sum_{m=-\infty}^{\infty} \sum_{n=-\infty}^{\infty} g\left(\frac{m}{2\Omega_x}, \frac{n}{2\Omega_y}\right) \mathrm{sinc}(2\Omega_x x - m)\mathrm{sinc}(2\Omega_y y - n). \tag{2.10}$$

Here the sinc function is defined as:

$$\mathrm{sinc}(x) = \frac{\sin(\pi x)}{\pi x}. \tag{2.11}$$

A band-limited function is therefore typically expected to be well represented by $(4\Omega_x\Omega_y)(4L_xL_y)$ samples in an interval of size $(2L_x \times 2L_y)$. This number is referred to as the space-bandwidth product or degrees of freedom in a band-limited signal. For two-dimensional signals or images, the space-bandwidth product may be considered the same as the number of pixels required to represent the image. The sampling expansion is an orthogonal series representation of the band-limited signal, since the shifted sinc functions have the following orthogonality property:

$$\int_{-\infty}^{\infty} dx \ \mathrm{sinc}(2\Omega_x x - m) \ \mathrm{sinc}(2\Omega_x x - n) = \frac{1}{2\Omega_x}\delta_{m,n}. \tag{2.12}$$

Here m and n are integers and $\delta_{m,n}$ is the Kronecker delta symbol which is equal to 1 for $m = n$ and zero otherwise. In optical systems the wave-field functions are typically band-limited due to the system aperture which limits the spatial frequencies passed by the system. The sampling theorem is therefore very useful in deciding the sampling requirements associated with wave fields.

2.2 Review of random processes theory

The study of light propagation through turbulence requires us to deal with the phenomenon of random time-varying fluctuations in the local refractive index of air. The properties of a well-defined beam propagating in a vacuum are essentially deterministic. For given initial amplitude, phase, and polarization profiles of a spatially coherent beam, its characteristics after propagation in a vacuum over a certain distance can be readily estimated. Refractive index of the atmosphere is close to that of a vacuum but is not a constant. Even small random fluctuations in the local refractive index of the order of 10^{-5}–10^{-6} can significantly disturb the beam characteristics after it has propagated over a distance of a few kilometers. Further, if the light beam under study is partially coherent, the electric field is not deterministic and shows inherent space–time fluctuations that need to be treated as a random process. Overall, the ideas of the random process theory are therefore important for us throughout this book.

The numerical values assumed by a variable u representing a physical quantity as a function of some continuous parameter t (e.g. time) may not be predictable. They may, however, follow a probability distribution function $p(u, t)$. At a given t, integrating over all possible values that may be taken by u, we may write:

$$\int_{-\infty}^{\infty} du\ p(u,\ t) = 1. \tag{2.13}$$

We may imagine an ensemble of processes $u^{(1)}(t)$, $u^{(2)}(t)$, … , $u^{(M)}(t)$ derived out of the same probability distribution $p(u,\ t)$. Any theoretical treatment of the quantity u and description of what is typically observed in an experiment will necessarily have to be in terms of averages over this ensemble. Any conclusions based on the behavior of individual realizations of the process may not be meaningful. The expected mean of u for example may be evaluated as:

$$\langle u\rangle = \int_{-\infty}^{\infty} du\ p(u,\ t)u. \tag{2.14}$$

The notation $\langle\cdots\rangle$ above denotes ensemble average. While the probability density $p(u,\ t)$ provides statistical information about u at t, additional interesting information about u can be obtained by knowing about its correlations at t_1 and t_2. This correlation information is contained in the joint probability density $p(u_1,\ t_1;\ u_2,\ t_2)$ which is proportional to the probability that u takes the values over a small interval centered on u_1 at t_1 and u_2 at t_2. In principle, joint densities of all higher order correlations are required for the complete description of a random process. In practice the most important correlation used is of second order and is denoted as:

$$\Gamma(t_1,\ t_2) = \langle u^*(t_1)\ u(t_2)\rangle. \tag{2.15}$$

The complex conjugate on the first term above makes the definition Hermitian symmetric:

$$\Gamma(t_1,\ t_2) = \Gamma^*(t_2,\ t_1). \tag{2.16}$$

Also, when $t_1 = t_2$, this quantity simply reduces to the expected intensity $\langle |u\ |^2\rangle$ of the underlying field. A random process $u(t)$ is called wide sense stationary if both $p(u,\ t)$ and the joint probability density $p(u_1,\ t_1;\ u_2,\ t_2)$ have no preferred origin in time t. The correlation function $\Gamma(t_1,\ t_2)$ in such a case is only a function of the time difference $(t_2 - t_1)$. A random process is called ergodic if the expectation values associated with the process can be evaluated using a long time average over its single realization. For example, for an ergodic process $u(t)$,

$$\langle u\rangle = \lim_{T\to\infty}\frac{1}{T}\int_{-T/2}^{T/2} dt\ u^{(n)}(t). \tag{2.17}$$

A sufficient condition for ergodicity to hold is that the correlation function $\Gamma(\tau = t_2 - t_1)$ dies out to zero sufficiently fast for large τ. In this case a single realization of the process can be divided into uncorrelated intervals that may themselves be considered as separate realizations of the random process. An important result related to the correlation function that we will have occasion to use is the Wiener–Khintchine theorem [3, 4] which states that the spectral density or spectrum of the random process is related to the correlation function by a Fourier transform relation.

$$S(\nu) = \int_{-\infty}^{\infty} d\tau\ \Gamma(\tau)\exp(-i2\pi\nu\tau). \tag{2.18}$$

We define the truncated Fourier transform of the random process $u(t)$ as:

$$U(\nu, T) = \int_{-T/2}^{T/2} dt\ u(t)\exp(-i2\pi\nu t). \tag{2.19}$$

The result in equation (2.18) can be proved by ensemble averaging the periodogram $S_T(\nu)$ defined as

$$S_T(\nu) = \frac{|U(\nu, T)|^2}{T} \tag{2.20}$$

in the long time limit. In other words, it can be shown that

$$S(\nu) = \lim_{T\to\infty}\left\langle \frac{|U(\nu, T)|^2}{T} \right\rangle. \tag{2.21}$$

It is easy to understand the Wiener–Khintchine relation intuitively. If we associate $u(t)$ with the electric field of light, it is well known for example, that for a narrow-band source, the correlation $\Gamma(\tau)$ survives longer, whereas for a broadband source the correlation dies down fast. In fact, the extent over which $|\Gamma(\tau)|$ remains significant is termed as the coherence time τ_c. In describing refractive index fluctuations in the atmosphere, we will be concerned with a space-domain random process $n(\mathbf{r})$, which represents the refractive index of the atmosphere at location $\mathbf{r}$. If the correlation function of the process defined as

$$\Gamma_n(\mathbf{r}_1, \mathbf{r}_2) = \langle n(\mathbf{r}_1)\, n(\mathbf{r}_2)\rangle, \tag{2.22}$$

is dependent only on the difference $(\mathbf{r}_1 - \mathbf{r}_2)$, then such a process is called a homogeneous process (analogous to stationary process). In that case a Wiener–Khintchine spectrum for $\Gamma_n(\mathbf{r}_1 - \mathbf{r}_2)$ may be defined in a manner similar to equation (2.18), which plays an important role in modeling of atmospheric turbulence, as we shall see later in this book.

A random process may be represented over a finite interval $[-T/2, T/2]$ by means of an orthogonal basis representation.

$$u(t) = \sum_{n=0}^{\infty} a_n\, \psi_n(t), \tag{2.23}$$

where

$$\int_{-\infty}^{\infty} dt\ \psi_n^*(t)\,\psi_m(t) = \delta_{m,n}. \tag{2.24}$$

There is a multiplicity of choices for the orthogonal basis set to be used for representing the random process. However, in order to make this representation efficient, we desire that the individual basis functions capture the essential independent features of the random process. The coefficients a_n should therefore be uncorrelated and should satisfy the following property in an ensemble averaged sense.

$$\langle a_m^* a_n \rangle = \lambda_n \, \delta_{m,n} \tag{2.25}$$

for positive constants λ_n. The auto-correlation function of the process can be written in terms of the basis representation as follows:

$$\Gamma(t_1, t_2) = \langle u^*(t_1)\, u(t_2) \rangle = \sum_{n=0}^{\infty} \lambda_n \, \psi_n^*(t_1)\, \psi_n(t_2). \tag{2.26}$$

Further, the basis functions are eigenfunctions associated with the integral kernel $\Gamma(t_1, t_2)$:

$$\lambda_n \psi_n(t_2) = \int_{-T/2}^{T/2} dt_1 \, \Gamma(t_1, t_2)\, \psi_n(t_1). \tag{2.27}$$

It is possible to show that the basis function set defined as above is most efficient in the sense that it captures maximal energy in $u(t)$ for a given finite number of terms in the orthogonal series representation in equation (2.23). This best basis representation is known by the name Kosambi–Karhunen–Loeve (KKL) expansion [5–7]. The discrete version of this best basis representation is known as the principal component analysis (PCA). PCA is an important statistical tool for unsupervised classification of datasets in diverse application areas.

2.2.1 Principal component analysis (PCA)

The random processes in practice represent discretely sampled data arrays. For example, we may consider digital images of human faces as a random process represented by pixel values in the face images. Individual elements of such data vectors will vary in detail but are expected to have a similar underlying structure. PCA serves as an effective method for uncovering significant patterns and extracting relevant information from such high-dimensional datasets [8–11]. Initially rooted in multivariate data analysis, PCA is applied across diverse fields, such as signal de-noising, neuroscience, financial services, population genetics, microbiome studies, texture analysis, atmospheric sciences, data compression and notably, machine learning algorithms. In machine learning, PCA is widely used as a preliminary step in the analysis of large datasets as PCA reduces the dimensionality of data, allowing algorithms to process and analyze information more efficiently. PCA reduces the dimensionality of large datasets through a vector space transformation. Typically, these datasets are burdened with potentially correlated variables, noise and extraneous information. PCA facilitates the transformation of these datasets by projecting them onto a new basis, known as principal components (PCs). The main aim here is, to represent the high-dimensional data using a minimum number of PCs. This transformation helps to filter out the noise and retain the most significant features of the data, enhancing the overall quality and interpretability of the dataset. This simplification makes it easier to detect trends, patterns, and outliers in the data.

In the ensuing discussion, we shall explore the PCA technique in terms of its linear algebra fundamentals. Consider an $m \times n$ matrix $\boldsymbol{D}$ representing the original dataset,

$$\boldsymbol{D} = \begin{bmatrix} x_{11} & x_{12} & \cdots & x_{1n} \\ x_{21} & x_{22} & \cdots & x_{2n} \\ \vdots & \vdots & \ddots & \vdots \\ x_{m1} & x_{m2} & \cdots & x_{mn} \end{bmatrix} = \begin{bmatrix} r_1 \\ r_2 \\ \vdots \\ r_m \end{bmatrix} \tag{2.28}$$

where the m rows correspond to the measured features or variables, and the n columns represent the measurements. The aim of PCA is to determine whether a new representation of the dataset, as an $m \times n$ matrix $\boldsymbol{Z}$, can be found, which is formed by linear combinations of the original basis and optimally re-expresses the data. Therefore, we write the equation for change of basis as [11],

$$\boldsymbol{Z} = \boldsymbol{PD} \tag{2.29}$$

where $\boldsymbol{P}$ is the matrix that transforms $\boldsymbol{D}$ to $\boldsymbol{Z}$. The rows of $\boldsymbol{P}$ give the desired new set of basis vectors. Consider the rows of $\boldsymbol{P}$ as the row vectors $\boldsymbol{p}_1, \boldsymbol{p}_2, \ldots, \boldsymbol{p}_m$, and the columns of $\boldsymbol{D}$ to be the column vectors $\boldsymbol{c}_1, \boldsymbol{c}_2, \ldots, \boldsymbol{c}_n$, then

$$\boldsymbol{PD} = \begin{bmatrix} \boldsymbol{p}_1 \\ \boldsymbol{p}_2 \\ \vdots \\ \boldsymbol{p}_m \end{bmatrix} \begin{bmatrix} \boldsymbol{c}_1 & \boldsymbol{c}_2 & \cdots & \boldsymbol{c}_n \end{bmatrix} \tag{2.30}$$

Therefore, we can write that

$$\boldsymbol{Z} = \begin{bmatrix} \boldsymbol{p}_1\boldsymbol{c}_1 & \boldsymbol{p}_1\boldsymbol{c}_2 & \cdots & \boldsymbol{p}_1\boldsymbol{c}_n \\ \boldsymbol{p}_2\boldsymbol{c}_1 & \boldsymbol{p}_2\boldsymbol{c}_2 & \cdots & \boldsymbol{p}_2\boldsymbol{c}_n \\ \vdots & \vdots & \ddots & \vdots \\ \boldsymbol{p}_m\boldsymbol{c}_1 & \boldsymbol{p}_m\boldsymbol{c}_2 & \cdots & \boldsymbol{p}_m\boldsymbol{c}_n \end{bmatrix} \tag{2.31}$$

The data $\boldsymbol{D}$ is projected onto the columns of $\boldsymbol{P}$, where the rows of $\boldsymbol{P}$, $\{\boldsymbol{p}_1, \boldsymbol{p}_2, \ldots, \boldsymbol{p}_m\}$, provide a new basis for representing $\boldsymbol{D}$ [12]. These rows of $\boldsymbol{P}$ define the directions of the PCs.

In order to identify the noise and redundancy in the original dataset, PCA uses the symmetric $m \times m$ covariance matrix $\boldsymbol{C_D}$ of $\boldsymbol{D}$,

$$\boldsymbol{C_D} = \frac{1}{n-1}\boldsymbol{D}\boldsymbol{D}^T \tag{2.32}$$

In general, covariance measures the extent of the linear relationship between two variables. A large positive covariance indicates that the data is positively correlated, while a large negative covariance signifies negative correlation. The absolute magnitude of the covariance reflects the degree of redundancy between the variables (see figure 2.1). The diagonal elements of the covariance matrix $\boldsymbol{C_D}$ represent the variances of the features, while the off-diagonal elements indicate the correlations between all possible pairs of features. High values in the off-diagonal terms suggest greater redundancy in the data. In order to reduce the redundancy in the data, the

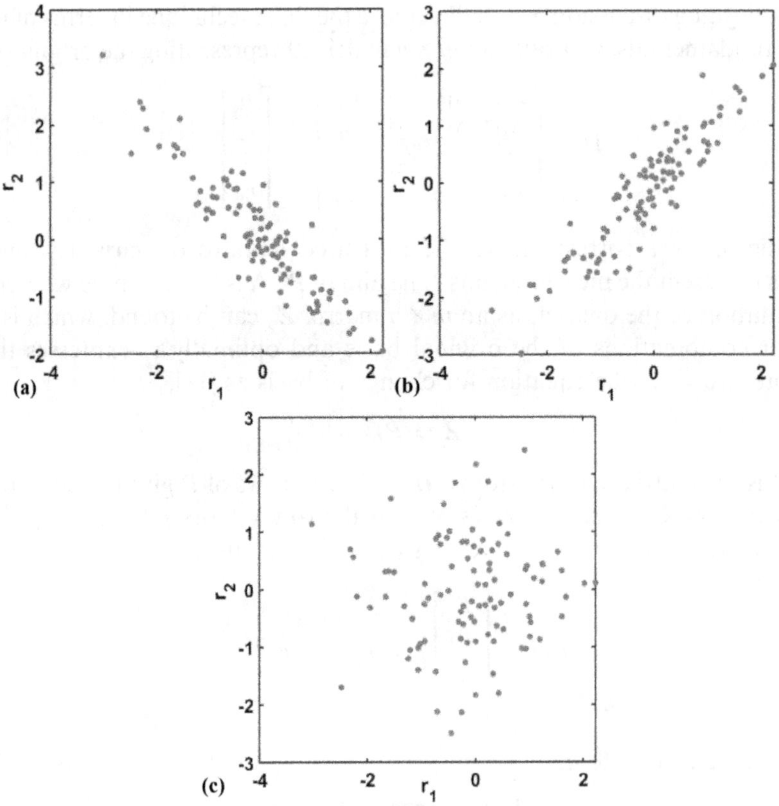

Figure 2.1. An example illustrating the spectrum of possible redundancies in data from two separate measurements, r_1 and r_2. The measurements in (a) and (b) are highly negatively and positively correlated, respectively, indicating a high level of redundancy. In contrast, (c) shows uncorrelated measurements, where one cannot predict one from the other.

PCA method assumes that the new representation of the dataset, given by $\boldsymbol{Z}$ should be as uncorrelated as possible. Therefore, the off-diagonal elements in the covariance matrix $\boldsymbol{C_Z}$ of $\boldsymbol{Z}$, given by

$$\boldsymbol{C_Z} = \frac{1}{n-1}\boldsymbol{Z}\boldsymbol{Z}^T \tag{2.33}$$

should be minimized. In other words, if some transformation matrix $\boldsymbol{P}$ exists which will produce a diagonal $\boldsymbol{C_Z}$ matrix, then the redundancy in the data can be removed. The PCA method operates under two important assumptions: (1) PCs are orthogonal, allowing for the use of linear algebra decomposition techniques, and (2) larger variances typically indicate significant features of the data, whereas smaller variances may indicate noise. One algebraic solution to PCA can be obtained using the eigenvector decomposition.

$$\begin{aligned} C_Z &= \frac{1}{n-1} ZZ^T \\ C_Z &= \frac{1}{n-1} PD(PD)^T \\ C_Z &= \frac{1}{n-1} PDD^T P^T \end{aligned} \tag{2.34}$$

Using equation (2.32), we obtain

$$C_Z = PC_D P^T \tag{2.35}$$

To proceed, we note that C_D is an $m \times m$ symmetric matrix. According to the principles of linear algebra, every square symmetric matrix is orthogonally diagonalizable [13]. Thus, we can write

$$C_D = EAE^T \tag{2.36}$$

where A is a diagonal matrix which has eigenvalues of C_D as its diagonal entries and E is an orthonormal matrix of orthonormal eigenvectors of A arranged as columns. Therefore, if we choose P to be a matrix where each row p_i is an eigenvector of C_D, then from equation (2.35), $P = E^T$ and

$$C_D = P^T AP \tag{2.37}$$

Substituting this in equation (2.35) gives,

$$\begin{aligned} C_Z &= P(P^T AP)P^T \\ C_Z &= (PP^T)A(PP^T) \end{aligned} \tag{2.38}$$

We know that the inverse of an orthogonal matrix is its transpose, $P^T = P^{-1}$, and $PP^{-1} = I$, where I is the identity matrix, so we can write

$$\begin{aligned} C_Z &= (PP^{-1})A(PP^{-1}) \\ C_Z &= A \end{aligned} \tag{2.39}$$

Clearly, selecting this form for matrix P achieves the diagonalization of C_Z, which was the main objective of PCA. An important aspect of PCA is that it automatically determines the relative importance of each PC through their variances. This is achieved during the diagonalization process. After calculating the eigenvalues and eigenvectors of C_D, the eigenvalues are sorted in descending order and placed on the diagonal of A. The orthonormal matrix P is then constructed by arranging the corresponding eigenvectors in the same order, forming the rows of P. The eigenvector associated with the largest eigenvalue is placed in the first row, the one with the second largest in the second row, and so forth. The component with the largest variance is called the first PC, second largest the second PC, and so on. In practice, before computing PCA, the dataset D needs to be normalized so as to have zero mean and unit standard deviation. The MATLAB environment has an in-built *pca* function [14] for obtaining the PCs.

2.3 Simulating a random process with known spectral density

The standard method for numerically generating a realization of a random process $u(t)$ with a given spectral density $S(\nu)$, is the Fourier transform method. For example, a random realization $u^k(t)$ can be generated by taking inverse Fourier transform of the quantity $\sqrt{S(\nu)} \times \exp[i\theta^k(\nu)]$. Here $\theta^k(\nu)$ is a uniform or normally distributed random phase map. We observe that the ensemble averaged periodogram corresponding to $u^k(t)$ now automatically yields the desired spectral density $S(\nu)$ by design. As an illustration, we show three realizations of a random process described by the triangular spectrum in figures 2.2(a)–(c). Here the spectral density is defined as:

$$S(\nu) = \Lambda\left(\frac{\nu}{\Delta\nu}\right), \tag{2.40}$$

with $\Lambda(x) = 1 - |x|$ for $|x| \leqslant 1$ and equal to zero everywhere else. For the illustrations, we have used the numerical value $\Delta\nu = 6$. Figure 2.2(d) shows average Fourier power spectrum computed over 500 realizations of the random process. We observe that the shape of average spectrum resembles that of the triangle function as expected.

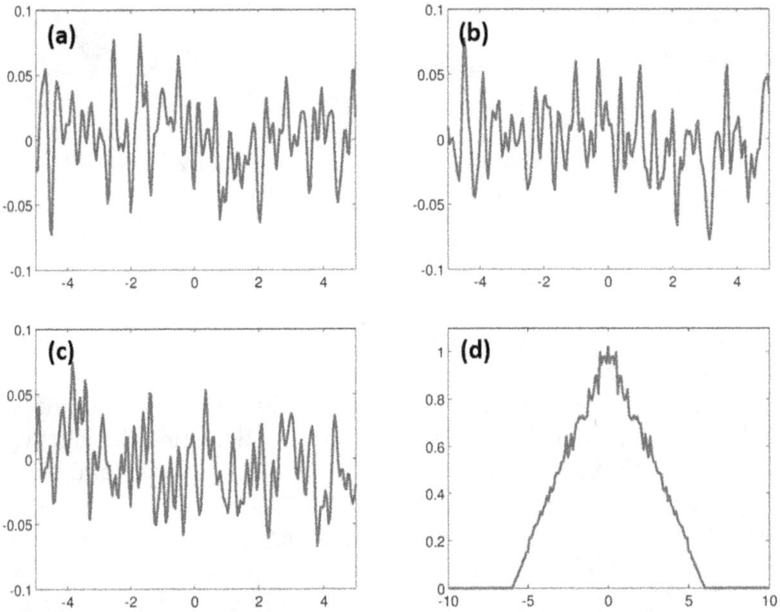

Figure 2.2. Panels (a)–(c) show realizations of a random process with power spectrum $S(\nu)$ given by the triangle function defined in equation (2.40) with $\Delta\nu = 6$. Only the real part of the random process is shown here. Panel (d) illustrates the average of Fourier transform magnitude squares calculated over 500 realizations of the random process. The y-axis units in panel (d) are arbitrary. The approximate triangular shape of the power spectrum can be clearly observed.

2.4 Complex signal representation

Complex representation of real valued random processes is commonly used in description of optical fields, coherence theory and in optical signal processing. We will examine this concept in detail in this section. The Fourier transform $U(\nu)$ of a real valued signal $u_r(t)$ has the symmetry property:

$$U(\nu) = U^*(-\nu). \tag{2.41}$$

As a result, the negative frequency components do not contain any additional information compared to the positive frequency components. The negative frequencies may, therefore, be omitted without any loss of information. This may be achieved by multiplying the Fourier transform $U(\nu)$ with a unit step function. The corresponding time domain signal is referred to as the complex or analytic signal representation as first suggested by Gabor [15]. The complex signal $u_z(t)$ can therefore be represented as:

$$u_z(t) = \frac{1}{2}\int_{-\infty}^{\infty} d\nu \; U(\nu) \; [1 + \mathrm{sgn}(\nu)] \; \exp(i2\pi\nu t). \tag{2.42}$$

Here sgn(ν) is the signum function defined as:

$$\begin{aligned} \mathrm{sgn}(\nu) &= \;\;\, 1 \;\text{ for }\; \nu > 0 \\ &= -1 \;\text{ for }\; \nu < 0. \end{aligned} \tag{2.43}$$

The term $[1 + \mathrm{sgn}(\nu)]/2$ in the above expression represents the unit step function. The sgn filter in frequency domain as in equation (2.42) corresponds to Hilbert transform in the real (or t) space and the complex signal can be expressed as:

$$u_z(t) = \frac{1}{2}u(t) + \frac{i}{2\pi}\mathrm{P}\int_{-\infty}^{\infty} dt' \; \frac{u(t')}{t - t'}. \tag{2.44}$$

The symbol 'P' here denotes the principal value of the integral. For a narrow-band optical field with central frequency ν_0, the real valued field is typically represented as

$$u_r(t) = a(t) \cos[2\pi\nu_0 t + \Phi(t)], \tag{2.45}$$

where, $a(t)$ and $\Phi(t)$ are amplitude and phase functions that vary slowly compared to the carrier frequency signal at frequency ν_0. If $u_r(t)$ represents scalar field associated with realistic laboratory sources, $a(t)$ and $\Phi(t)$ are not deterministic and have to be treated as random processes. Using the relations in equations (2.42) and (2.44), it can be shown that:

$$\begin{aligned} u_z(t) &= \frac{1}{2}a(t)(\cos[2\pi\nu_0 t + \Phi(t)] + i\,\sin[2\pi\nu_0 t + \Phi(t)]) \\ &= \frac{1}{2}a(t)\exp[i2\pi\nu_0 t + i\Phi(t)]. \end{aligned} \tag{2.46}$$

We observe that in equation (2.45), we are representing a single function $u_r(t)$ in terms of two functions $a(t)$ and $\Phi(t)$. Such a representation is not therefore unique and an additional criterion is required to make an appropriate choice of the two functions $a(t)$ and $\Phi(t)$. The complex representation as obtained above using the Hilbert transform relation in equation (2.44) was obtained by suppressing the redundant information. One may ask if this representation is optimal in some sense. This question was addressed by Mandel in an interesting early work [16] which for some reason is not widely known in the optics as well as signal processing community.

Mandel's result is as follows. Suppose $u_r(t)$ represents a real valued random process. We will consider another random process generated by convolving $u_r(t)$ with a filter $k(t)$:

$$u_r'(t) = \int_{-\infty}^{\infty} dt'\, k(t - t')\, u_r(t), \tag{2.47}$$

and a construct a complex signal:

$$\begin{aligned} v_z(t) &= u_r(t) + i\, u_r'(t) \\ &= u_{zo}(t)\exp(i2\pi\nu_0 t). \end{aligned} \tag{2.48}$$

For a stationary random process $u_r(t)$, the power spectral density is $S_r(\nu)$ (defined as per the Wiener–Khintchine theorem equation (2.18)) and the power spectral density $S_z(\nu)$ for the process $v_z(t)$ may be related by:

$$S_z(\nu) = S_r(\nu)|1 + iK(\nu)|^2. \tag{2.49}$$

Mandel's approach was to seek an appropriate filter $k(t)$ such that the fluctuation in the complex envelope $u_{zo}(t)$ defined by:

$$(Du_z)^2 = \left\langle \left| \frac{du_{zo}(t)}{dt} \right|^2 \right\rangle \tag{2.50}$$

is minimized. The solution of this minimization problem can be obtained analytically by methods of variational calculus. The minimization of the quantity $(Du_z)^2$ in equation (2.50) can be shown to be equivalent to the solution of the following constrained optimization problem:

$$\text{minimize} \int_{-\infty}^{\infty} d\nu (\nu - \nu_0)^2 S_z(\nu) \tag{2.51}$$

with the constraint that

$$\int_{-\infty}^{\infty} d\nu (\nu - \nu_0) S_z(\nu) = 0. \tag{2.52}$$

The solution of this minimization problem as shown by Mandel is that the filter $k(t)$ is the Hilbert transform that occurs in the second term of equation (2.44). Since Hilbert transform connects cosines and sines (in one dimension), Mandel's result indicates that the cosine–sine quadrature representation, which is commonly used for optical fields, is particularly special because it provides the most efficient, or least fluctuating, representation for the complex envelope of the field.

2.5 Spiral phase quadrature transform

The cosine–sine quadrature representation in one dimension and its specialty in the sense of Mandel's theorem leads us to seek similar representation in higher dimensions. In this section, we will extend Mandel's idea to two-dimensional signals in order to determine a transform equivalent to the Hilber transform. The treatment presented here, essentially follows the work described in [17]. Later in this book we will see that the notion of quadrature transform in two dimensions as presented here has important implications for designing laser beams that can maintain robust intensity profiles on passing through turbulence.

We consider a two-dimensional signal $g(r, \theta)$ which is a realization of a stationary (or homogeneous) random field. We will assume the mean $\langle g(r, \theta)\rangle$ to be zero. The stationarity implies that the correlation function $\Gamma(\mathbf{r}_1, \mathbf{r}_2) = \langle g^*(\mathbf{r}_1)g(\mathbf{r}_2)\rangle$ is only a function of the difference $(\mathbf{r}_1 - \mathbf{r}_2)$ and the power spectral density $S(\rho)$ is related to the correlation function $\Gamma(\mathbf{r}_1 - \mathbf{r}_2)$ via the Wiener–Khintchine theorem. The spectral density $S(\rho)$ of a real valued process $g(\mathbf{r})$ is expected to have inversion symmetry:

$$S(\rho, \phi) = S(\rho, \phi + \pi). \tag{2.53}$$

In analogy with the one-dimensional case, we seek a filter function $K(\rho, \phi)$ that will be a generalization of the $\mathrm{sgn}(\nu)$ filter corresponding to the Hilbert transform and define the complex signal as:

$$g_z(r, \theta) = \frac{1}{2}\int\int \rho\, d\rho\, d\phi\; G(\rho, \phi)[1 + iK(\rho, \phi)]\exp[i2\pi r\rho\cos(\theta - \phi)]. \tag{2.54}$$

The spectral density of $g_z(r, \theta)$ denoted by $S_z(\rho, \phi)$ can thus be written as:

$$S_z(\rho, \phi) = \frac{1}{4}S(\rho, \phi)|1 + i\,K(\rho, \phi)|^2. \tag{2.55}$$

Since the $\mathrm{sgn}(\nu)$ filter in one dimension preserves the magnitude of the signal at all non-zero frequencies, we will further look for a unit magnitude filter of the form:

$$K(\rho, \phi) = \exp[i\alpha(\rho, \phi)], \tag{2.56}$$

with the requirement

$$K(\rho, \phi + \pi) = -K(\rho, \phi), \tag{2.57}$$

so that, the two-dimensional definition of the quadrature transform remains consistent with the one-dimensional definition. Next we define a complex envelope $a(r, \theta)$ corresponding to the signal $g_z(r, \theta)$:

$$a(r, \theta) = \frac{1}{2}\int\int \rho\, d\rho\, d\phi\; G(\rho, \phi)[1 + i\,K(\rho, \phi)]\exp[i2\pi r(\rho - \rho_0)\cos(\theta - \phi)]. \tag{2.58}$$

Here ρ_0 is the carrier frequency of the signal $g_z(r, \theta)$ defined as:

$$\rho_0 = \frac{\int\int \rho d\rho\, d\phi\ \ [\rho\ S_z(\rho, \phi)]}{\int\int \rho d\rho\, d\phi\ [S_z(\rho, \phi)]} \tag{2.59}$$

The fluctuation in the envelope function analogous to equation (2.50) for this two-dimensional case may be defined using the Parseval theorem:

$$\begin{aligned} \left\langle \left| \frac{\partial a(r, \theta)}{\partial r} \right|^2 + \frac{1}{r^2} \left| \frac{\partial a(r, \theta)}{\partial \theta} \right|^2 \right\rangle &= \pi^2 \int\int \rho d\rho d\phi\ (\rho - \rho_0)^2 S(\rho, \phi)|1 + iK(\rho, \phi)|^2 \\ &= 4\pi^2 \int\int \rho d\rho d\phi\ (\rho - \rho_0)^2 S_z(\rho, \phi). \end{aligned} \tag{2.60}$$

The task of finding the appropriate filter function thus reduces to the following variational optimization problem:

$$\text{minimize} \int\int \rho d\rho d\phi\ (\rho - \rho_0)^2\ S_z(\rho, \phi), \tag{2.61}$$

subject to the constraint:

$$\int\int \rho d\rho d\phi\ (\rho - \rho_0)\ S_z(\rho, \phi) = 0. \tag{2.62}$$

The variational optimization problem is therefore essentially the same as that described in equations (2.51) and (2.52), respectively, for the one-dimensional case. Note that we have assumed the filter $K(\rho, \phi)$ to be in the form $\exp[i\alpha(\rho, \phi)]$, so that,

$$|1 + iK(\rho, \phi)|^2 = 2[1 - \sin[\alpha(\rho, \phi)]. \tag{2.63}$$

Assuming the solution to the minimization problem in equation (2.61) to be $\alpha(\rho, \phi)$, if we vary the solution to the neighboring function $\alpha(\rho, \phi) + \varepsilon\eta(\rho, \phi)$ for small positive number ε, then the integral in equation (2.61) must remain unchanged to first order in ε. In performing this variation the function $\eta(\rho, \phi)$ only needs to satisfy the condition in equation (2.57):

$$\exp[i\alpha(\rho, \phi) + i\varepsilon\ \eta(\rho, \phi)] = -\exp[i\alpha(\rho, \phi + \pi) + i\varepsilon\ \eta(\rho, \phi + \pi)]. \tag{2.64}$$

which is already satisfied when $\varepsilon = 0$. Therefore, $\eta(\rho, \phi)$ is arbitrary but needs to satisfy the symmetry condition:

$$\eta(\rho, \phi) = \eta(\rho, \phi + \pi). \tag{2.65}$$

The variation in $\alpha(\rho, \phi)$ also changes the mean frequency ρ_0 to $(\rho_0 - \delta\rho_0)$ with $\delta\rho_0$ proportional to ε. Making these substitutions in equation (2.61), we can collect the terms that are first order in ε and set their sum to zero.

$$\begin{aligned} &2\delta\rho_0 \int\int \rho d\rho d\phi\ (\rho - \rho_0)S(\rho, \phi)\{1 - \sin[\alpha(\rho, \phi)]\} \\ &- \varepsilon \int\int \rho d\rho d\phi\ (\rho - \rho_0)^2\ S(\rho, \phi)\eta(\rho, \phi)\cos[\alpha(\rho, \phi)] = 0. \end{aligned} \tag{2.66}$$

The first term above vanishes due to the constraint in equation (2.62). In view of the symmetry properties of $S(\rho, \phi)$, $K(\rho, \phi)$ and $\eta(\rho, \phi)$, the integral in the second term can be rewritten as:

$$\int_0^{\pi} d\phi \int \rho d\rho \ (\rho - \rho_0)^2 S(\rho, \phi)\eta(\rho, \phi)\{\cos[\alpha(\rho, \phi)] + \cos[\alpha(\rho, \phi + \pi)]\} = 0. \quad (2.67)$$

In order for the above integral to vanish identically for arbitrary functions $S(\rho, \phi)$ and $\eta(\rho, \phi)$, we must have:

$$\cos[\alpha(\rho, \phi)] + \cos[\alpha(\rho, \phi + \pi)] = 0. \quad (2.68)$$

We observe that the simplest choice of such a function that satisfies this condition is:

$$\alpha(\rho, \phi) = \phi. \quad (2.69)$$

The filter corresponding to this simple choice is:

$$K(\rho, \phi) = \exp(i\phi). \quad (2.70)$$

The filter is thus defined everywhere except at $\rho = 0$ where ϕ is not defined. This should not be a problem for us since $g(r, \theta)$ is assumed to be a zero mean process which has zero dc component. We therefore have an important result, that the spiral phase filter $K(\rho, \phi) = \exp(i\phi)$ defined for two-dimensional signals is analogous to the sgn(ν) filter in one-dimension in the sense of Mandel's theorem. The spiral phase filtering operation can thus be thought of as the quadrature transform for two-dimensional signals. The work of K G Larkin and co-workers [18] has already suggested through a number of numerical experiments that the spiral phase filter indeed is a suitable extension of the concept of Hilbert transform to two-dimensional signals. It is important to note that the Bessel functions of first kind $J_0(\alpha r)$ and $J_1(\alpha r)$ with $r = \sqrt{x^2 + y^2}$ are exactly related by the spiral phase transform apart from some constant multipliers.

As an illustration of the spiral phase transform, we consider the two-dimensional low-frequency band represented by circle function:

$$\mathrm{circ}\left(\frac{\rho}{a}\right) = 1 \quad \text{for} \quad \rho \leqslant a$$
$$= 0 \quad \text{otherwise}. \quad (2.71)$$

Here $\rho = \sqrt{f_x^2 + f_y^2}$ is the radial coordinate in the two-dimensional Fourier space. The two functions

$$g_1(r, \theta) = \mathcal{F}^{-1}[\mathrm{circ}(\rho/a)], \quad (2.72)$$

and

$$g_2(r, \theta) = \mathcal{F}^{-1}[\exp(i\phi)\ \mathrm{circ}(\rho/a)]. \quad (2.73)$$

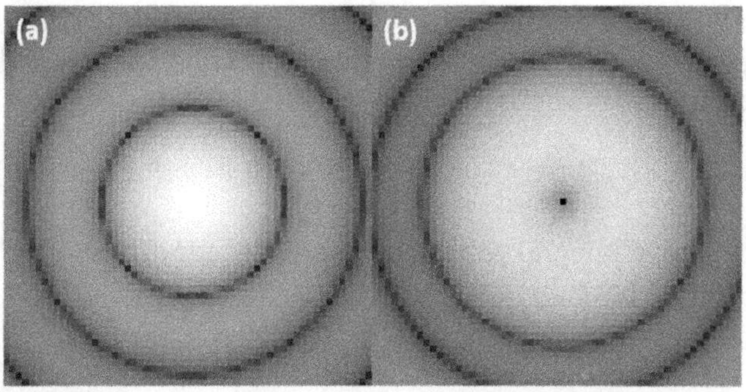

Figure 2.3. Illustration of complementary signals produced by the spiral phase transform. Panels (a) and (b) show central portions (57 × 57 pixels) of $|g_1(r, \theta)|$ and $|g_2(r, \theta)|$ as defined in equations (2.72) and (2.73), respectively. The complementarity in the magnitude patterns can be seen visually.

with $\phi = \arctan(f_y/f_x)$ should therefore form a 'two-dimensional quadrature' pair of functions showing sine–cosine like diversity. For numerical computation, we define the circ function over a 256 × 256 pixel array and choose $a = 8$. The central 57 × 57 portions of the Fourier transform magnitude images associated with $|g_1(r, \theta)|$ and $|g_2(r, \theta)|$ defined above are shown in figure 2.3. We observe that the locations of maxima and minima in the two images in figure 2.3 are indeed exchanged suggesting sine–cosine-like diversity in the functions $|g_1(r, \theta)|$ and $|g_2(r, \theta)|$.

The spiral phase transform is closely related to the orbital angular momentum states of light that are central to the discussion in this book. We will show later that the complementarity above holds for general amplitude-phase apertures. In this context, the result will also be found to be useful when we discuss engineering of laser beams that show robust intensity profile on propagation through turbulence.

References

[1] Bracewell R N 1987 *The Fourier Transform and Its Applications* (New York: McGraw-Hill)

[2] Osgood B G Ee261: the Fourier transform and its applications, Stanford engineering online course material available at https://see.stanford.edu/course/ee261.

[3] Mandel L and Wolf E 1995 *Optical Coherence and Quantum Optics* (Cambridge: Cambridge University Press)

[4] Goodman J W 2015 *Statistical Optics* 2nd edn (New York: Wiley)

[5] Kosambi D D 1943 Statistics in function space *J. Indian Math. Soc.* **7** 76–88

[6] Karhunen K 1947 Uber lineare methoden in der wahrscheinlichkeitsrechnung *Ann. Acad. Sci. Fennicae. Ser. A I. Math.-Phys.* **37** 1–79

[7] Loeve M 1948 Fonctions alatoires du second ordre *Processus Stochastique et Mouvement Brownien* ; P Levy

[8] Abdi H and Williams L J 2010 Principal component analysis *WIREs Comput. Stat.* **2** 433–59

[9] Wold S, Esbensen K and Geladi P 1987 Principal component analysis *Chemometr. Intell. Lab. Syst.* **2** 37–52 (Proceedings of the Multivariate Statistical Workshop for Geologists and Geochemists)

[10] Radhakrishna Rao C 1964 The use and interpretation of principal component analysis in applied research *Sankhyā: Ind. J. Stat. Ser. A (1961-2002)* **26** 329–58

[11] Shlens J 2014 A tutorial on principal component analysis or arXiv:1404.1100v1

[12] Richardson M 2009 Principal component analysis (Duke University) http://people.maths.ox.ac.uk/richardsonm/SignalProcPCA.pdf

[13] Strang G 2023 *Introduction to Linear Algebra* (Cambridge: Wellesley)

[14] MathWorks Principal component analysis of raw data https://in.mathworks.com/help/stats/pca.html

[15] Gabor D 1946 Theory of communications part 1: the analysis of information *Trans. Inst. Electr. Eng.* **3** 429–56

[16] Mandel L 1967 Complex representation of optical fields in coherence theory *J. Opt. Soc. Am.* **57** 613–7

[17] Khare K 2008 Complex signal representation, Mandel's theorem, and spiral phase quadrature transform *Appl. Opt.* **47** E8–12

[18] Larkin K G, Bone D J and Oldfield M A 2001 Natural demodulation of two-dimensional fringe patterns. i. General background of the spiral phase quadrature transform *J. Opt. Soc. Am.* A **18** 1862–70

IOP Publishing

Orbital Angular Momentum States of Light (Second Edition)
Propagation through atmospheric turbulence
Kedar Khare, Priyanka Lochab and Paramasivam Senthilkumaran

Chapter 3

The angular spectrum method

Propagation of wave fields is of basic importance to this book and we will discuss the angular spectrum approach [1] in this context. While our main concern is propagation of light through turbulence, it is important to lay out the methodology for propagation of fields in free space. Later, we will see that the typical methodology for modeling atmospheric propagation involves propagating wave fields through free space interspersed with random phase screens, where the angular spectrum method will be found to be useful.

3.1 Wave equation

The wave equation is obtained from the Maxwell's equations and describes the propagation of light through a medium. For a general medium the Maxwell equations are given by:

$$\nabla \times \boldsymbol{H} = \frac{\partial \boldsymbol{D}}{\partial t} + \boldsymbol{J} \tag{3.1}$$

$$\nabla \times \boldsymbol{E} = -\frac{\partial \boldsymbol{B}}{\partial t} \tag{3.2}$$

$$\nabla .\ \boldsymbol{D} = \rho \tag{3.3}$$

$$\nabla .\ \boldsymbol{B} = 0 \tag{3.4}$$

Here $\boldsymbol{H}, \boldsymbol{E}, \boldsymbol{D}$ and $\boldsymbol{B}$ represent magnetic field, electric field, electric displacement and the magnetic induction, respectively. The symbol ρ stands for the volume free charge density and $\boldsymbol{J}$ gives the electric current density. The magnetic field $\boldsymbol{H}$ and the electric displacement $\boldsymbol{D}$ are related to $\boldsymbol{B}$ and $\boldsymbol{E}$ through the constitutive relations. In a linear, isotropic medium $\boldsymbol{H}$ and $\boldsymbol{D}$ are related to $\boldsymbol{B}$ and $\boldsymbol{E}$ as:

doi:10.1088/978-0-7503-5959-7ch3

$$\boldsymbol{H} = \frac{\boldsymbol{B}}{\mu} \tag{3.5}$$

$$\boldsymbol{D} = \varepsilon \boldsymbol{E} \tag{3.6}$$

where μ is permeability and ε is the permittivity of the medium. For free space, the permeability and dielectric constant have values

$$\mu_o = 4\pi \times 10^{-7}\ \text{H m}^{-1} \tag{3.7}$$

$$\varepsilon_o = 8.854 \times 10^{-12}\ \text{F m}^{-1} \tag{3.8}$$

and the speed of light in free space is given by:

$$c = \frac{1}{\sqrt{\mu_o \varepsilon_o}} = 3 \times 10^8\ \text{m s}^{-1}. \tag{3.9}$$

The permeability and permittivity of a medium are related to their vacuum values through the relations:

$$\mu = \mu_r \mu_o \tag{3.10}$$

$$\varepsilon = \varepsilon_r \varepsilon_o \tag{3.11}$$

where μ_r and ε_r are known as the relative permeability and relative permittivity of the medium. The dielectric constant can also be written in terms of the refractive index n as:

$$\varepsilon = \varepsilon_o n^2 \tag{3.12}$$

Let us assume that the medium has constant magnetic permeability and zero conductivity. Taking the curl of equation (3.2) and rearranging the partial derivative and curl operator on the right hand side, we get:

$$\nabla \times (\nabla \times \boldsymbol{E}) = -\frac{\partial}{\partial t}(\nabla \times \boldsymbol{B}) \tag{3.13}$$

The double curl operation on the left hand side can be simplified using the vector identity:

$$\nabla \times (\nabla \times \boldsymbol{E}) = \nabla(\nabla.\ \boldsymbol{E}) - \nabla^2 \boldsymbol{E} \tag{3.14}$$

Substituting $\boldsymbol{H}$ and $\boldsymbol{D}$ in terms of $\boldsymbol{B}$ and $\boldsymbol{E}$ in equations (3.1) and (3.3) leads to:

$$\nabla \times \boldsymbol{B} = \mu \frac{\partial}{\partial t}(\varepsilon \boldsymbol{E}) \tag{3.15}$$

$$\nabla.\ (\varepsilon \boldsymbol{E}) = 0 \tag{3.16}$$

We further evaluate equation (3.16) by using the vector identity:

$$\nabla.\ (u\boldsymbol{A}) = u(\nabla.\ \boldsymbol{A}) + \boldsymbol{A}.\ (\nabla u)$$

which gives us;

$$\nabla.\ (\varepsilon \boldsymbol{E}) = \varepsilon(\nabla.\ \boldsymbol{E}) + \boldsymbol{E}.\ (\nabla\varepsilon) = 0 \tag{3.17}$$

or,

$$\nabla.\ \boldsymbol{E} = -\boldsymbol{E}.\ \frac{(\nabla\varepsilon)}{\varepsilon} = -\boldsymbol{E}.\ \nabla \ln \varepsilon \tag{3.18}$$

This allows us to write equation (3.13) in the form:

$$\nabla^2 \boldsymbol{E} + \nabla(\boldsymbol{E}.\ \nabla \ln \varepsilon) = \mu \frac{\partial^2}{\partial t^2}(\varepsilon \boldsymbol{E}) \tag{3.19}$$

The second term on the left hand side gives the *depolarization of light* and represents the coupling between the orthogonal polarizations of the electric field on propagation through a medium which has space-dependent permittivity [2]. For a medium with constant ε, for example free space, the term $\nabla(\boldsymbol{E}.\ \nabla \ln \varepsilon) = 0$. We can then also write

$$\mu \frac{\partial^2}{\partial t^2}(\varepsilon \boldsymbol{E}) = \mu\varepsilon \frac{\partial^2 \boldsymbol{E}}{\partial t^2}$$

These simplifications along with equations (3.9)–(3.12) give us the final form of the wave propagation equation in a medium of refractive index n as:

$$\nabla^2 \boldsymbol{E}(x, y, z, t) - \frac{n^2}{c^2}\frac{\partial^2}{\partial t^2}\boldsymbol{E}(x, y, z, t) = 0 \tag{3.20}$$

A similar wave equation can also be written for the magnetic field $\boldsymbol{H}$:

$$\nabla^2 \boldsymbol{H}(x, y, z, t) - \frac{n^2}{c^2}\frac{\partial^2}{\partial t^2}\boldsymbol{H}(x, y, z, t) = 0 \tag{3.21}$$

Since the same vector equation is satisfied by both $\boldsymbol{E}$ and $\boldsymbol{H}$, it means that all components of these vectors also satisfy an identical scalar equation. Therefore, we can for the moment ignore the vectorial nature of the wave equation and work with scalar quantities. Let the scalar optical field be given by $\tilde{U}(x, y, z, t)$, then

$$\nabla^2 \tilde{U}(x, y, z, t) - \frac{n^2}{c^2}\frac{\partial^2}{\partial t^2}\tilde{U}(x, y, z, t) = 0 \tag{3.22}$$

For monochromatic waves, the time-dependence of the optical fields is sinusoidal of the form $\exp(-i\omega t)$ where ω is the frequency of the wave,

$$\tilde{U}(x, y, z, t) = u(x, y, z)\exp(-i\omega t). \tag{3.23}$$

Substituting this form of $\tilde{U}(x, y, z, t)$ into equation (3.22), one gets the time-independent form of the scalar wave equation, also known as the *Helmholtz wave equation*,

$$(\nabla^2 + k^2)u(x, y, z) = 0 \tag{3.24}$$

where $k = \omega/c = 2\pi/\lambda$ is the wavenumber and λ is the wavelength.

3.2 The angular spectrum formalism

The problem of interest to us involves solving the monochromatic wave equation or the Helmholtz equation:

$$(\nabla^2 + k^2)\, u(x, y; z) = 0, \tag{3.25}$$

between two planes transverse to the nominal propagation direction ($+z$). Given the wave-field $u(x, y; 0)$ in the $z = 0$ plane, we would like to determine the field in the right half space ($z > 0$). The Helmholtz equation in free-space may be readily analyzed using a Fourier transform based method as we will discuss here. In later chapters of this book we will present the more involved case when the medium of propagation is filled with turbulent air with space-time varying refractive index. We begin by representing the field $u(x, y; z)$ as a two-dimensional Fourier expansion with respect to the transverse coordinates (x, y):

$$u(x, y; z) = \int\int df_x\, df_y \quad U(f_x, f_y; z) \exp[i2\pi(f_x x + f_y y)]. \tag{3.26}$$

Applying the operator $(\nabla^2 + k^2)$ on both sides of the above equation gives:

$$\int\int df_x df_y \left\{\frac{\partial^2}{\partial z^2} + [k^2 - 4\pi^2(f_x^2 + f_y^2)]\right\} U(f_x, f_y; z) \exp[i2\pi(f_x x + f_y y)] = 0. \tag{3.27}$$

In the above equation, we have used the derivative identity for the Fourier transform:

$$\frac{\partial u}{\partial x} = \int\int df_x\, df_y\, (2\pi i f_x) U(f_x, f_y; z) \exp[i2\pi(f_x x + f_y y)]. \tag{3.28}$$

A similar relation can be written for the y derivative. The z-coordinate is being treated separately from the transverse coordinates as z is the nominal direction in which we wish to propagate the field. Since the right hand side of equation (3.27) is identically zero for an arbitrary function $U(f_x, f_y, z)$, the only way in which the integral can be zero is if the integrand itself is equal to zero. Therefore we have:

$$\left[\frac{\partial^2}{\partial z^2} + \alpha^2\right] U(f_x, f_y; z) = 0, \tag{3.29}$$

where we have introduced the notation:

$$\alpha^2 = k^2 - 4\pi^2(f_x^2 + f_y^2). \tag{3.30}$$

The solution of the equation (3.29) can be readily written as:

$$U(f_x, f_y; z) = A(f_x, f_y)\exp(i\alpha z) + B(f_x, f_y)\exp(-i\alpha z). \tag{3.31}$$

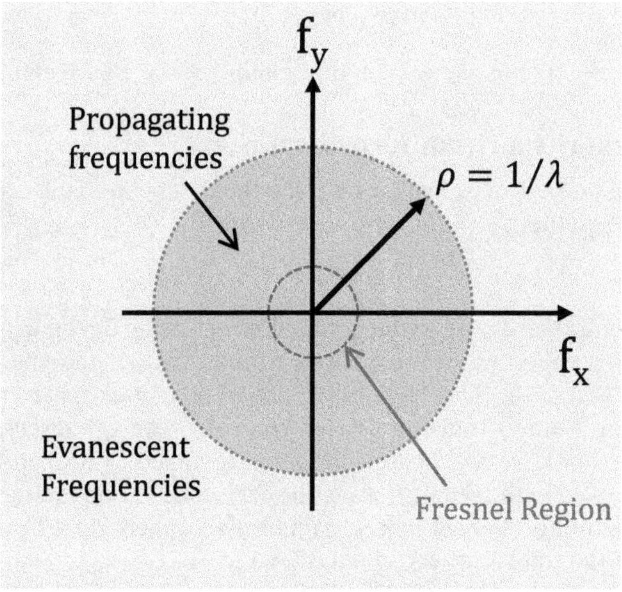

Figure 3.1. The division of spatial frequency plane into propagating and evanescent components is illustrated. The Fresnel zone corresponds to paraxial frequency components close to the origin of the spatial frequency plane.

We note that in the $\exp(-i\omega t)$ convention the first and second terms above represent the outgoing and the incoming solutions, respectively. We further note that the spatial frequency plane may be divided into two distinct regions as shown in figure 3.1.

$$f_x^2 + f_y^2 \leqslant \frac{1}{\lambda^2}, \tag{3.32}$$

where $\alpha^2 \geqslant 0$ and the region

$$f_x^2 + f_y^2 > \frac{1}{\lambda^2}, \tag{3.33}$$

where α is an imaginary quantity. For spatial frequencies outside the circle of radius $1/\lambda$ centered on origin of the two-dimensional Fourier space, the first term in equation (3.31) therefore decays exponentially, whereas the second term grows exponentially. Since we are interested in the outgoing solution which needs to satisfy the boundary conditions at $z \to \infty$, we set $B(f_x, f_y) = 0$. Further since the field in the $z = 0$ plane is already known, on setting $B(f_x, f_y) = 0$, it is easy to see that

$$A(f_x, f_y) = U(f_x, f_y; 0). \tag{3.34}$$

The two-dimensional Fourier transform of the diffracted field after propagation by distance z can therefore be represented as:

$$U(f_x, f_y; z) = U(f_x, f_y; 0)\exp(i\alpha z). \tag{3.35}$$

Here it is to be noted that for the spatial frequencies (f_x, f_y) satisfying the condition in equation (3.32), the quantity $\exp(i\alpha z)$ is purely a phase factor. The spatial frequencies satisfying the condition in equation (3.33) are seen to decay exponentially with z. The spatial frequency components within and outside the circle of radius $1/\lambda$ in the two-dimensional Fourier space are therefore called propagating and evanescent components, respectively [3, 4]. In the present book, we will mostly not have an occasion to deal with the evanescent components as the propagation distances of interest are in kilometers. The field $u(x, y, z)$ can be evaluated by inverse transforming the Fourier space relation in equation (3.35):

$$u(x, y, z) = \int\int df_x\, df_y\, [U(f_x, f_y; 0)\exp(i\alpha z)]\exp[i2\pi(f_x x + f_y y)]. \tag{3.36}$$

We observe that the propagating part of diffracted field can be decomposed as a linear combination of plane waves propagating in the directions represented by the vector $(f_x, f_y, \alpha/(2\pi))$ and having amplitudes $[U(f_x, f_y, 0)\exp(i\alpha z)]$. This methodology of treating the diffraction problem is therefore referred to as the angular spectrum approach. We observe that for the propagating components, the angle made by the k-vector of the plane wave with x, y, z axes is $\theta_x = \arccos(\lambda f_x)$, $\theta_y = \arccos(\lambda f_y)$ and $\theta_z = \sqrt{1 - \cos^2(\theta_x) - \cos^2(\theta_y)}$, respectively. The low spatial frequency region near $(f_x, f_y) = (0, 0)$ thus corresponds to plane waves with wave vector close to the z-axis (paraxial case). For the limiting spatial frequencies satisfying the relation $f_x^2 + f_y^2 = (1/\lambda^2)$, the corresponding plane waves have the wave vector in the x–y plane. The product relation in two-dimensional Fourier space as in equation (3.35) corresponds to a convolution in (x, y) space and is given by the well-known first Rayleigh–Sommerfeld–Smythe diffraction formula [5]:

$$u(x, y; z) = \int\int dx'\, dy'\, u(x', y'; 0)\, \frac{\exp(ikR)}{2\pi R}\left(-ik + \frac{1}{R}\right)\frac{z}{R}. \tag{3.37}$$

Here $R = \sqrt{(x - x')^2 + (y - y')^2 + z^2}$ is the distance between the points $(x', y', 0)$ and (x, y, z). The relation in equation (3.37) is consistent with the Maxwell equations. It is easy to show that this relation reduces to the Fresnel and Fraunhofer relations in the appropriate paraxial limits. For example, for optical wavelengths and typical laboratory distances $|-ik| > >1/R$ and for paraxial angles $z/R \approx 1$. This leads to the wide angle form of Fresnel diffraction:

$$u(x, y; z) = -ik \int\int dx'\, dy'\, u(x', y'; 0)\, \frac{\exp(ikR)}{2\pi R}. \tag{3.38}$$

When R is further approximated assuming $z > >(x - x')$ and $z > >(y - y')$ and in particular when

$$z^3 > > \frac{k}{8}[(x - x')^2 + (y - y')^2]_{\max}, \tag{3.39}$$

the above expression reduces to the familiar paraxial form [3]:

$$u(x, y; z) = \frac{-ike^{ikz}}{2\pi R} \int\int dx' \, dy' \, u(x', y'; 0) \, e^{\frac{i\pi}{\lambda z}[(x-x')^2 + (y-y')^2]}. \tag{3.40}$$

Finally, when $z > > \frac{k(x'^2 + y'^2)_{\max}}{2}$, the Fresnel approximation formula above can be simplified further leading to the Fraunhofer approximation [3]:

$$u(x, y; z) = \frac{-ike^{ikz} \, e^{\frac{i\pi}{\lambda z}(x^2 + y^2)}}{2\pi R} \int\int dx' \, dy' \, u(x', y'; 0) \, e^{-i\frac{2\pi}{\lambda z}(xx' + yy')}. \tag{3.41}$$

Note that the Fraunhofer diffraction formula above shows that the far-field diffraction pattern for a finite aperture illuminated by a plane wave is nothing but the two-dimensional Fourier transform of the field in the $z = 0$ plane.

3.3 Sampling considerations and usage of fast Fourier transform routines

For general beam profile structures, it is easier to implement the free-space propagation calculation using the relation equation (3.35). The required forward and inverse Fourier transforms can be evaluated using the fast Fourier transform (FFT) routines that are readily available with standard programming tools. While propagating the field over a distance z it is important to pay attention to sampling considerations. If the computational window size is $L \times L$, then over a distance z the spatial frequencies involved are in the range:

$$\Delta f = \pm \frac{\sin \theta_{\max}}{\lambda} = \pm \frac{(L/2)}{\lambda \sqrt{z^2 + (L/2)^2}}. \tag{3.42}$$

Here $\theta_{\max}$ is the angle made by a line joining the on-axis point in the $z = 0$ plane with the edge of the computational window located at the plane $z = L$. As a result, when sampling the field at discrete points $(m\Delta x, n\Delta y)$ for integers m and n, the sampling intervals must be selected to satisfy the sampling criterion:

$$\Delta x, \quad \Delta y \leqslant \frac{1}{2\Delta f}. \tag{3.43}$$

Numerical computation of field propagation can be performed easily using Fourier transform based approach as is evident from equation (3.35). Here we describe some finer points associated with usage of the FFT routines readily available in popular computational programs such as MATLAB, Octave, Python, etc for the purpose of field propagation calculations. We note that the transfer function relation allows us to write the field $u(x, y; z)$ as:

$$u(x, y; z) = \mathcal{F}^{-1}\{\mathcal{F}[u(x, y; 0)]\exp(i\alpha z)\}. \tag{3.44}$$

Suppose that the field in the $z = 0$ plane is sampled over a window of $L \times L$ and the appropriate sampling criterion as in equation (3.43) has been met. For a 1-dimensional function $u(m\Delta x)$ with $N = (L)/(\Delta x)$ samples, the FFT routine leads to the computation:

$$U(n) = \sum_{m=0}^{N-1} u(m\Delta x)\exp(-i2\pi mn/N), \quad n = 0, 1, 2, \dots, (N-1). \tag{3.45}$$

The standard function call: $U = fft(u)$ performs the above numerical computation and leads to a new vector U with the same length N. The spatial frequencies associated with the elements of this vector U are ordered as:

$$0, \frac{1}{L}, \frac{2}{L}, \dots, \frac{N-1}{L}.$$

Note that the first element of the vector U is associated with the zero frequency. Secondly note that the zero phase factor is applied to the first element $u(m = 0)$ of the input function. In a two-dimensional version of the function call (the `fft2` function), the zero frequency will be located at the left corner of the result matrix. For the optical field applications it is more appropriate that the zero phase be applied to the on-axis element, for example in accordance with the Fraunhofer diffraction formula in equation (3.41). A simple trick to make sure we get numerical results consistent with what is expected in optics laboratory experiments is to use the function call `fft` as follows: `U` = `fftshift(fft(ifftshift` `(u)))`. The function *ifftshift* shifts the central 'on-axis' element of the vector u above to the first place by swapping the two halves of the vector. The function `fftshift` on the other hand readjusts the result of `fft` function call such that the zero frequency component is brought back on-axis to the central element. The two functions `fftshift` and `ifftshift` are identical when N is even and differ by a single element shift when N is odd. The function call suitable for inverse FFT is similar and is given by: u = `fftshift(ifft(ifftshift(U)))`. Note that here `fft` has been replaced by `ifft` (inverse FFT), however, the sequence of `fftshift` and `ifftshift` has remained the same. We would like to emphasize here that the use of `fftshift` or `ifftshift` functions is only required here in order to obtain amplitude and phase profiles of the field that are consistent with what is expected in a laboratory experiment. This discussion is important for practical implementation of field propagation, otherwise, it is possible to get unphysical computational results (particularly for the phase map of the resultant fields).

3.4 Numerical propagation of fields in free space

As an illustration, we show the free-space propagation of a vortex beam described in the $z = 0$ plane as:

$$u(r, \theta, z = 0) = Ar\exp(-r^2/W_o^2)\exp(i\theta), \tag{3.46}$$

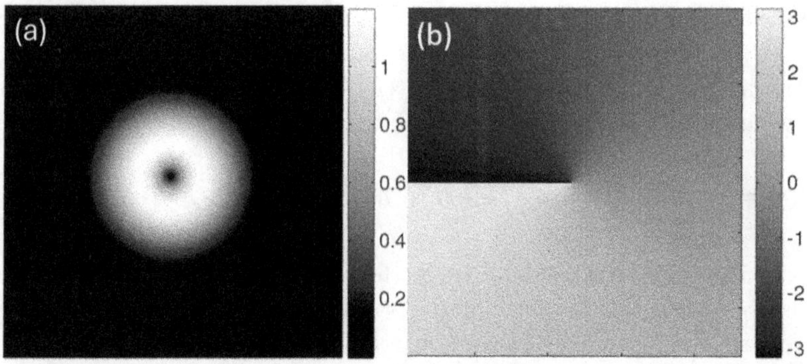

Figure 3.2. The amplitude and phase profile of the field described in equation (3.46) are shown in (a) and (b), respectively.

where $r = \sqrt{x^2 + y^2}$ and $\theta = \arctan(y/x)$. As we will explain later in the book, this profile corresponds to a vortex beam in the orbital angular momentum state $l = 1$. The term 'vortex' here is suggestive of the phase singularity at $r = 0$ where phase is undefined. The amplitude and phase profile of the beam in $z = 0$ plane is shown in figures 3.2(a) and (b) respectively. The beam waist parameter W_0 is taken to be 100λ. The propagation distance is set to $z = 100\lambda$, 100.25λ, and 100.5λ, respectively. For these propagation distances, the angular spectrum over the computational window is essentially purely propagating in nature. As per sampling criterion in equation (3.43), a square grid of $M \times M$ samples is required where $M = 1001$. The amplitude of the field at $z = 100\lambda$ and the phase of the propagated field at the three distances are shown in figures 3.3(a)–(d), respectively. The amplitude of the field at $z = 100.25\lambda$ and 100.5λ is not shown separately as it is similar to that shown in figure 3.3(a). It is important to note that the phase jump at $\theta = \pi$ in the phase profile shows rotation by 90 and 180 degrees for the distances $z = 100.25\lambda$ and $z = 100.5\lambda$, respectively. This is a well-known feature of a vortex beam that its phase jump (or phase cut) nominally rotates with a periodicity of λ. This is an interesting feature of the field that has naturally come out of the angular spectrum based propagation method.

While the numerical illustration shown here is approximately correct, we note that the input field in equation (3.46) changes very rapidly near the phase singularity at $r = 0$. The discrete sampling of the field at intervals $(\Delta x, \Delta y)$ is clearly not adequate to represent this fast varying field near the vortex core ($r = 0$) in this case. Examination of the field near the singularity needs additional discussion that will be provided in the next chapter. The periodicity ($= \lambda$) of the phase jump along the propagation distance z does not follow trivially from the Rayleigh–Sommerfeld diffraction formula and the discussion in the next chapter will show that this periodicity is generally observed only if we are not near the singularity at $r = 0$. These points however, demand a more careful semi-analytical analysis using the angular spectrum approach.

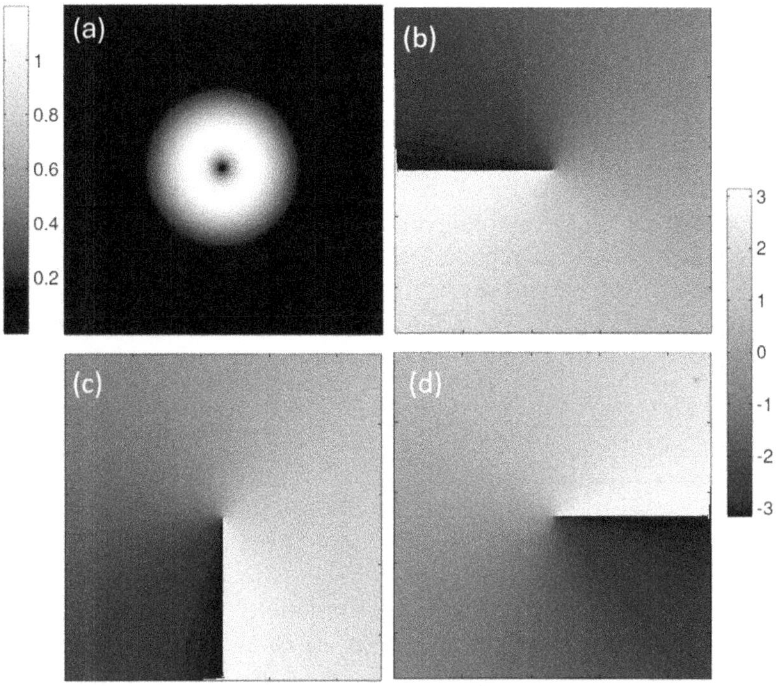

Figure 3.3. (a) The amplitude of the field (given in figure 3.2) after propagation to a distance of $z = 100\lambda$ is illustrated. Panels (b)–(d) depict the phase map of the propagated field at distances $z = 100\lambda$, 100.25λ, and 100.5λ, respectively.

3.5 Propagation of partially spatially coherent light fields

The treatment of free-space light propagation so far has been discussed for fully spatially coherent light fields. In practice, monochromatic light fields encountered in real-world applications may be spatially partially coherent. The partially coherent fields are characterized by the cross-spectral density $W(\boldsymbol{r}_1, \boldsymbol{r}_2, \nu)$ defined as a spatial correlation function:

$$W(\boldsymbol{r}_1, \boldsymbol{r}_2, \nu) = \langle u^*(\boldsymbol{r}_1, \nu)\, u(\boldsymbol{r}_2, \nu) \rangle. \tag{3.47}$$

Here the average $\langle \cdots \rangle$ is taken over the monochromatic ensemble following the theory of spatial coherence in the space-frequency domain [6]. The cross-spectral density is Hermitian in nature with respect to the variables $\boldsymbol{r}_1$ and $\boldsymbol{r}_2$, in other words,

$$W(\boldsymbol{r}_1, \boldsymbol{r}_2, \nu) = [W(\boldsymbol{r}_2, \boldsymbol{r}_1, \nu)]^*. \tag{3.48}$$

The normalized cross-spectral density $\mu(\boldsymbol{r}_1, \boldsymbol{r}_2, \nu)$ is defined as

$$\mu(\boldsymbol{r}_1, \boldsymbol{r}_2, \nu) = \frac{W(\boldsymbol{r}_1, \boldsymbol{r}_2, \nu)}{\sqrt{W(\boldsymbol{r}_1, \boldsymbol{r}_1, \nu)\, W(\boldsymbol{r}_2, \boldsymbol{r}_2, \nu)}}, \tag{3.49}$$

and has a magnitude in the range (0, 1), where the values 0 and 1 denote the fully spatially incoherent and fully spatially coherent cases, respectively. Since the monochromatic fields themselves satisfy the Helmholtz equation given in equation (3.25), the cross-spectral density can be shown to satisfy the wave equation with respect to both $\mathbf{r}_1$ and $\mathbf{r}_2$.

$$(\nabla_1^2 + k^2)(\nabla_2^2 + k^2)\, W(\mathbf{r}_1, \mathbf{r}_2, \nu) = 0. \tag{3.50}$$

This relation suggests that the cross-spectral density itself evolves like a wave which evolves as guided by the Helmholtz equation. Assuming a planar source and using the Rayleigh–Sommerfeld diffraction formula (given in equation (3.37)) with respect to both the spatial variables allows us to understand the evolution of cross-spectral density on propagation. Denoting the transverse coordinates in the source and observation plane as $(\mathbf{r}_1', \mathbf{r}_2')$, and $(\mathbf{r}_1, \mathbf{r}_2)$, respectively, we have:

$$W(\mathbf{r}_1, \mathbf{r}_2, \nu) = \frac{\exp\left[ik(R_2 - R_1)\right]}{4\pi^2 R_1 R_2} \iint_{\mathcal{A}} d^2\mathbf{r}_1' d^2\mathbf{r}_2'\, W(\mathbf{r}_1', \mathbf{r}_2', \nu)\left(ik + \frac{1}{R_1}\right)\left(-ik + \frac{1}{R_2}\right)\frac{z^2}{R_1 R_2}. \tag{3.51}$$

Here $R_j = \sqrt{z^2 + (x_j' - x_j)^2 + (y_j' - y_j)^2}$ for $j = 1, 2$ and the integrals are evaluated over the source area $\mathcal{A}$. When the source is spatially incoherent with source power spectral density $S_0(\mathbf{r}_1', \nu)$, we may write $W(\mathbf{r}_1', \mathbf{r}_2', \nu) = S_0(\mathbf{r}_1', \nu)\delta(\mathbf{r}_1' - \mathbf{r}_2')$. Further, if the observation plane is in the Fraunhofer zone relative to the source, then equation (3.51) for propagation of $W(\mathbf{r}_1', \mathbf{r}_2', \nu)$ in free-space reduces to

$$W(\mathbf{r}_1, \mathbf{r}_2, \nu) = \frac{\exp[ik(R_2 - R_1)]}{\lambda^2 R_1 R_2} \iint d^2\mathbf{r}_1'\; S_0(\mathbf{r}_1', \nu)\exp[-i2\pi\mathbf{r}_1' \cdot (\mathbf{r}_1 - \mathbf{r}_2)]. \tag{3.52}$$

This is the well-known van Cittert–Zernike theorem which states that the far-field cross-spectral density of a spatially incoherent source is proportional to the two-dimensional Fourier transform of the source power spectral density $S_0(\mathbf{r}_1', \nu)$. As per Wolf's space-frequency picture, the cross-spectral density of a partially coherent source can be expressed as

$$W(\mathbf{r}_1, \mathbf{r}_2, \nu) = \sum_n \sigma_n\, \phi_n^*(\mathbf{r}_1, \nu)\, \phi_n(\mathbf{r}_2, \nu). \tag{3.53}$$

Here, the set of functions $\{\phi_n(\mathbf{r}, \nu)\}$ are called the coherent modes of the source distributions. The coherent modes can be shown to be the eigenfunctions associated with $W(\mathbf{r}_1, \mathbf{r}_2, \nu)$ and satisfy:

$$\int d\mathbf{r}_1\; W(\mathbf{r}_1, \mathbf{r}_2, \nu)\, \phi_n(\mathbf{r}_1, \nu) = \sigma_n\, \phi_n(\mathbf{r}_2, \nu). \tag{3.54}$$

The cross-spectral density being Hermitian in nature, the functions $\{\phi_n(\mathbf{r}, \nu)\}$ are mutually orthogonal. One may appreciate that the treatment of spatially partially coherent light field has similarities to the KKL expansion of random processes as discussed in section 2.2. A partially coherent monochromatic source can be represented as

$$u(\mathbf{r}, \nu) = \sum_n a_n \phi_n(\mathbf{r}, \nu), \tag{3.55}$$

where the coefficients are uncorrelated and thus satisfying

$$\langle a_m^* a_n \rangle = |a_m|^2 \delta_{m,n}. \tag{3.56}$$

The picture of monochromatic partially spatially coherent light that emerges from this treatment, thus states that the field $u(\boldsymbol{r}, \nu)$ is an incoherent (or uncorrelated) sum of fully coherent modes. Such a light beam may be realized practically if we are able to combine radiation from multiple mutually incoherent sources (at the same wavelength) into a single beam. One possibility for example is to combine multiple independent lasers operating at nominally the same wavelength but individually engineered to operate in distinct spatial modes $\phi_n(\boldsymbol{r}, \nu)$. Another possibility is to use a temporally varying secondary source like a ground glass diffuser which is rotating (or moving) fast enough compared to detector response time, so that, the resultant irradiance at the detector consists of sum of large number of speckle realizations that are added incoherently. A simple and interesting realization of partially spatially coherent beam [7] proposed more recently uses a light emitting diode (LED) placed at the back focal plane of a lens. Since the LED active area has number of mutually incoherent point sources, the light beam beyond the lens aperture consists of incoherent combination of a number of plane waves with distinct k-vectors which serve as uncorrelated coherent modes. This partially coherent source has an interesting property that the spatial coherence length of such a combination is nearly propagation invariant as the individual coherent modes (plane waves) remain as plane waves even after arbitrary propagation distance.

The propagation of partially coherent light in free space (or a material medium) may therefore be understood as propagation of individual fully spatially coherent modes, whose propagated versions are be added incoherently to get an estimate of the ensemble averaged irradiance. The propagation of the individual fully spatially coherent modes $\phi_n(\boldsymbol{r}, \nu)$ in free space can be described using the angular spectrum approach as discussed earlier in section 3.2.

The cross-spectral density can be generalized further to include polarization. In that case the cross-spectral density $W(\boldsymbol{r}_1, \boldsymbol{r}_2, \nu)$ has multiple components and is defined as:

$$W_{jk}(\boldsymbol{r}_1, \boldsymbol{r}_2, \nu) = \langle u_j^*(\boldsymbol{r}_1, \nu)\, u_k(\boldsymbol{r}_2, \nu) \rangle. \tag{3.57}$$

Here the indices j, k represent any orthogonal polarization basis (for example, linear basis or circular basis). For the case when $\boldsymbol{r}_1 = \boldsymbol{r}_2$, we observe that $W_{jk}(\boldsymbol{r}_1, \boldsymbol{r}_2, \nu)$ is essentially the same as the polarization (or coherency) matrix used in definition of degree of polarization. The quantity $W_{jk}(\boldsymbol{r}_1, \boldsymbol{r}_2, \nu)$ is therefore, of central importance, for connecting coherence and polarization of light, which is an active research area on its own [8].

References

[1] Clemmow P C 1966 *The Plane Wave Spectrum Representation of Electromagnetic Fields* (Oxford: Pergamon)

[2] Tatarskii V I 1967 Depolarization of light by turbulent atmospheric inhomogeneities *Radiophys. Quantum Electron.* **10** 987–8
[3] Goodman J W 2016 *Introduction to Fourier Optics* 3rd edn (Boston, MA: Roberts)
[4] Khare K, Butola M and Rajora S 2023 *Fourier Optics and Computational Imaging* 2nd edn (Berlin: Springer)
[5] Mandel L and Wolf E 1995 *Optical Coherence and Quantum Optics* (Cambridge: Cambridge University Press)
[6] Wolf E 1982 New theory of partial coherence in the space–frequency domain. Part i: spectra and cross spectra of steady-state sources *J. Opt. Soc. Am.* **72** 343–51
[7] Aarav S, Bhattacharjee A, Wanare H and Jha A K 2017 Efficient generation of propagation-invariant spatially stationary partially coherent fields *Phys. Rev.* A **96** 033815
[8] Wolf E 2007 *Introduction to the Theory of Coherence and Polarization of Light* (Cambridge: Cambridge University Press)

Chapter 4

Near core structure of a propagating optical vortex

This chapter treats the problem of vortex propagation using a semi-analytical approach based on the angular spectrum method. The vortex phase field has a singularity at the center that is composed of high spatial frequency components. The vortex propagation problem cannot therefore be treated appropriately using the paraxial Fresnel approximation. Our analysis using the angular spectrum method reveals an interesting near-core structure for the propagating vortex. The result is seen to be a consequence of contribution from both propagating and evanescent spatial frequency components.

4.1 Vortex propagation using the angular spectrum method

In this chapter we present an illustration of the angular spectrum method for a propagating optical vortex and observe some interesting features of the vortex field in the near-core region. Optical vortices are commonly pictured as having a helical wavefront with periodicity of the helix equal to the wavelength λ. However, this periodicity of λ is somewhat puzzling, as there seems to be no simple explanation for this periodicity purely based on free-space diffraction. Propagation of vortex fields in free space thus demands additional investigation, as we describe in this chapter. The discussion in this chapter closely follows [1].

We begin by assuming an idealized initial scalar component of the vortex wave-field in the $z = 0$ plane given by:

$$u(r, \theta; 0) = \exp(i\theta), \tag{4.1}$$

where $\theta = \arctan(y/x)$ and $r = \sqrt{x^2 + y^2}$. The angle θ is not defined at $r = 0$ and the field will therefore have a null at this position. It is important to note that the input field has a singularity at the origin and as $r \to 0$ the input field changes

doi:10.1088/978-0-7503-5959-7ch4

arbitrarily fast implying that the input field $u(r, \theta; 0)$ has high spatial frequency content. Since Fresnel or Fraunhofer diffraction approximations are limited to the paraxial regime, they cannot account for the high spatial frequencies in the input field. We therefore examine propagation of the helical wavefront using the angular spectrum approach. The 2D Fourier transform of $u(r, \theta; 0) = \exp(i\theta)$ first needs to be evaluated in order to use the angular spectrum approach. The 2D Fourier transform of $\exp(i\theta)$ may be evaluated in polar coordinates as:

$$\int_0^{2\pi} d\theta \int_0^{\infty} r\, dr \exp(i\theta)\exp[-i2\pi r\rho\cos(\theta-\phi)] = -(2\pi i)\exp(i\phi)\int_0^{\infty} r dr J_1(2\pi r\rho). \quad (4.2)$$

Here $\phi = \arctan(f_y/f_x)$ is the polar angle in the two-dimensional spatial frequency plane and $\rho = \sqrt{f_x^2 + f_y^2}$. In the above equation, we have evaluated the integral over θ using the integral representation for Bessel function:

$$J_1(u) = \frac{1}{2\pi}\int_0^{2\pi} d\theta \exp(-i\theta)\exp(iu\sin\theta). \quad (4.3)$$

The integral of $r\ J_1(2\pi r\rho)$ on the right hand side of equation (4.2) can be evaluated using integration by parts where it will be handy to use the Bessel function relations:

$$\frac{d}{dx} J_0(x) = -J_1(x), \quad (4.4)$$

and

$$\int_0^{\infty} du\ J_0(u) = 1. \quad (4.5)$$

The two-dimensional Fourier transform of $u(x, y, 0)$ may be obtained using these Bessel function relations and is given by:

$$U(f_x, f_y; 0) = \frac{-i\exp(i\phi)}{2\pi\rho^2}. \quad (4.6)$$

A general form of this result for arbitrary order vortex beam can be derived using properties of Bessel functions and is given by [2]:

$$\int_0^{2\pi} d\theta \int_0^{\infty} r\, dr \exp(in\theta)\exp[-i2\pi r\rho\cos(\theta-\phi)] = \frac{|n|(-i)^{|n|}\exp(i\,n\,\phi)}{2\pi\rho^2} \quad (4.7)$$

As per the angular spectrum method, the field after propagating z distance, is obtained by inverse Fourier transforming $U(f_x, f_y; z)$ and may be expressed as:

$$u(r, \theta; z) = \int\int \rho d\rho\, d\phi\, \frac{-i\exp(i\phi)}{2\pi\rho^2}\exp(i\,z\sqrt{k^2-4\pi^2\rho^2})\exp[i2\pi r\rho\cos(\phi-\theta)], \quad (4.8)$$

where we have made use of equation (3.36) in polar coordinate form and equation (4.6). The integral over the polar angle ϕ can be readily evaluated giving:

$$u(r,\theta;z) = \exp(i\theta)\ V_1(r,z), \tag{4.9}$$

where

$$V_1(r,z) = \int_0^\infty d\rho\ \frac{\exp(i\,z\sqrt{k^2-4\pi^2\rho^2})}{\rho} J_1(2\pi r\rho). \tag{4.10}$$

It is to be noted that ρ in the denominator in the integrand above, does not pose a problem as $[J_1(2\pi r\rho)/\rho]$ is finite in the limit $\rho \to 0$. From equation (4.9) we note an interesting result that on starting with the vortex wavefront $\exp(i\theta)$, the resultant wavefront at any distance z is given by the same initial vortex function multiplied by $V_1(r,z)$ as in equation (4.10). It is important to note that the function $V_1(r,z)$ is identically equal to zero at $r=0$, since the Bessel function in the integrand vanishes at this point. The phase of the helical wavefront may now be written as:

$$\Phi(r,\theta;z) = \arg[u(r,\theta;z)] = \theta + \arg[V_1(r,z)]. \tag{4.11}$$

Contrary to the usual assumption about paraxial propagation of vortex wavefields, the phase of the helical wavefront is thus seen to depend on the radial coordinate r as well. In the previous chapter (figure 3.3), this behavior of the propagating vortex field was not evident due to the lack of sufficient sampling of the field in the input plane. It is clear that at the origin ($r=0$) both θ and $\arg[V_1(r,z)]$ are not defined and the propagated field continues to have null along the z-axis for arbitrary propagation distances.

4.2 Phase dip near vortex core

In this section, we will closely examine the near-core structure of a propagating optical vortex by numerically evaluating the results in equations (4.10) and (4.11). We have employed a standard numerical quadrature technique, in order to evaluate the integral in the definition of $V_1(r,z)$. The lower integration limit is $\rho = 0$ and the upper limit used for numerical integration is $\rho = 10/\lambda$, thus including evanescent components in the calculation. Our computation showed that including the evanescent spatial frequency components beyond this range did not make any significant difference to the numerical results. In figure 4.1, we plot the phase of the function $V_1(r,z)$ for $r = 0.1\lambda$ and $r = 1\lambda$ as z is varied from 0 to 1λ. The curves in figure 4.1 are plotted for the cases, when the upper limits of integration in equation (4.10) are set to $\rho = 0.5/\lambda$, $\rho = 1/\lambda$ and $\rho = 10/\lambda$. The plots in figure 4.1(a) show that as the limit of integration for ρ is increased, the phase of $V_1(r,z)$ shows a dip right after $z=0$. The black curve in this figure represents the numerical value of kz.

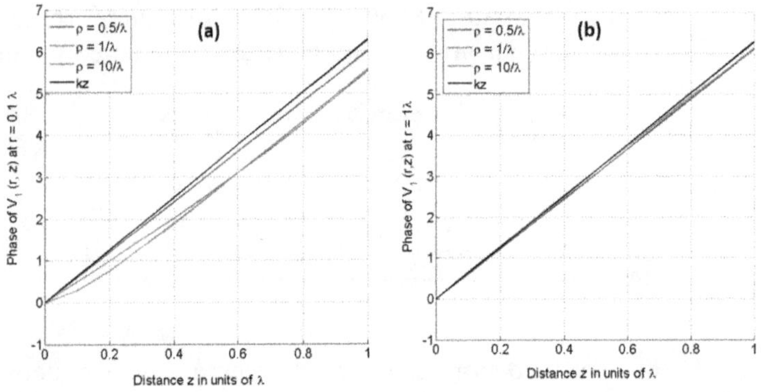

Figure 4.1. (a) and (b): Phase of function $V_1(r,\, z)$ near vortex core at $r = 0.1\lambda$ and $r = 1\lambda$, respectively. The z variation is kept close to the input plane in order to distinguish the different curves clearly. Adapted with permission from [1], copyright (2016) Optical Society of America.

The dip in phase significantly increases when evanescent frequencies are included in the computation compared to the case, when only propagating components (up to $\rho = 1/\lambda$) are included. The contribution of evanescent components, therefore seems to be important in this case in understanding the detailed near-core structure of a propagating vortex. The red curve representing the phase of $V_1(r,\, z)$ with inclusion of propagating as well as evanescent frequencies is seen to become parallel to the kz line but always lags behind it due to the initial phase dip. The case of $r = 1\lambda$ as shown in figure 4.1(b) shows that all the curves as in figure 4.1(a) are now close to the black line representing the quantity kz.

In figures 4.2(a), (c), and (e), the variation of phase of $V_1(r,\, z)$ as a function of r is shown at propagation distances of $z = 1\lambda$, 10λ and 100λ as two-dimensional surface plots. Figures 4.2(b), (d), and (f) show the total phase $\Phi(r,\, \theta,\, z)$ for the corresponding distances. These results indicate that there is a phase dip near the core of a propagating vortex beam. The magnitude of the phase dip is close to -0.8 radians and the physical extent of the phase dip grows as the beam propagates from $z = 0$. Additional investigation is needed to understand the magnitude of the phase dip near the core. In figures 4.2(b), (d), and (f) the units on x and y axes are in terms of wavelength λ. We observe that due to the phase dip, the contours of equal phase are no longer radial near the vortex core ($r = 0$) but are seen to spiral around the core. The presence of phase lag suggests that a propagating vortex does not have perfectly helical wavefront but near its core (region near $r = 0$) the constant phase contours show a spiralling effect. This phase lag effect is not fully studied in the literature. We believe that the angular spectrum method is a good tool to examine such effects as it can naturally handle both propagating and evanescent components in the field. Any additional ideas such as super-oscillatory functions to describe the fast varying field near the core are then not required. As seen in figure 4.2, as the vortex beam propagates, the phase lag near the core continues to exist and the region

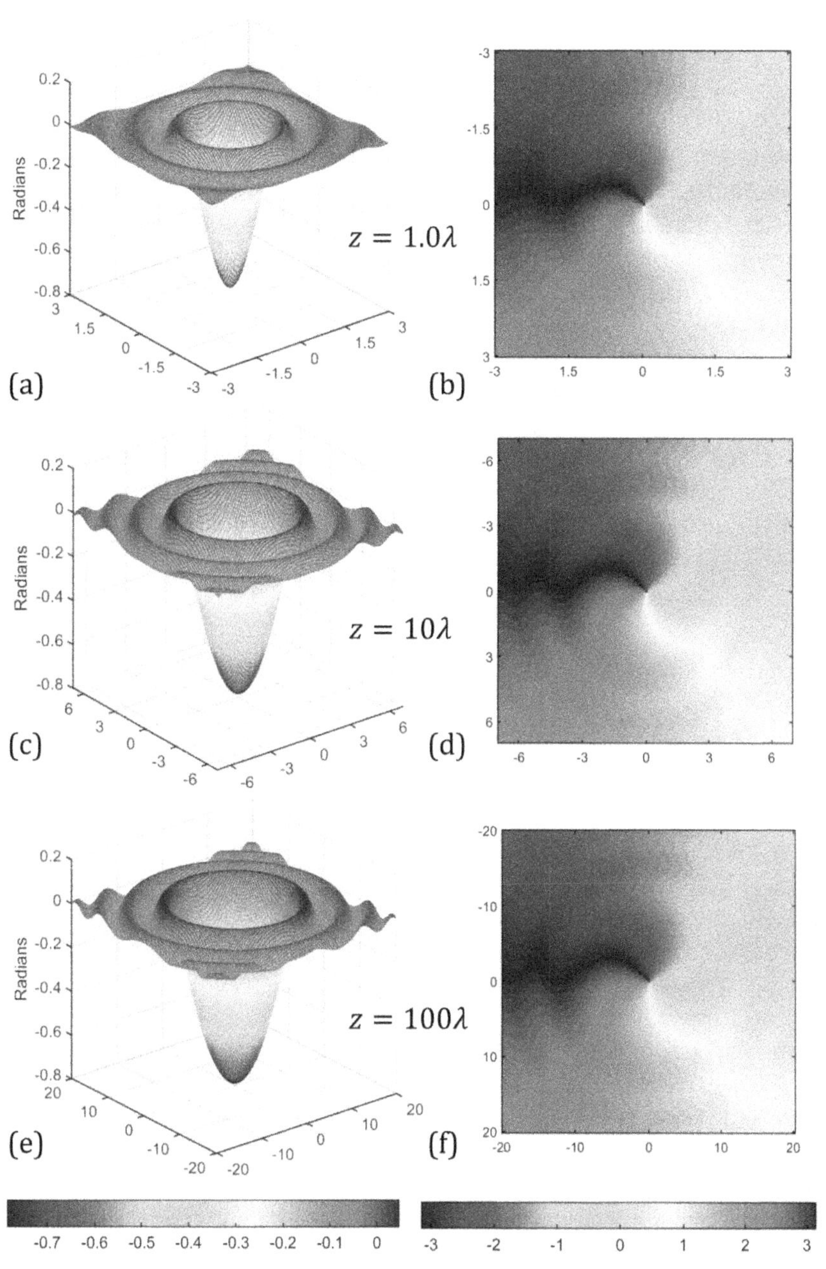

Figure 4.2. Panels (a), (c), and (e) display the retarded part of the wavefront at different propagation distances, while the corresponding total phase maps for $\Phi(r, \theta; z)$ are shown in (b), (d), and (f) respectively. The propagation distance z is 1λ for (a) and (b), 10λ for (c) and (d), and 100λ for (e) and (f). The range of the x–y coordinates in (a)and (b) spans from -3λ to 3λ, in (c) and (d) from -6λ to 6λ, and in (e) and (f) from -20λ to 20λ. Adapted with permission from [1], copyright (2016) Optical Society of America.

with spiral lines of phase contours grows due to free-space propagation. For a propagating phase front it is common to associate the **k**-vector associated with the wavefront, to the gradient $\nabla\Phi(r, \theta, z)$ of the wavefront phase function. Since the propagating vortex phase front has radial dependence, the transverse **k** -vector also has a non-zero radial component near vortex core. This term exists in addition to the usual azimuthal component due to the first term in equation (4.11). The effect of this radial component of **k**-vector has been observed experimentally [3] in the form of an anomalous fringe period in Young's double-slit experiment, where two closely spaced slits are illuminated symmetrically by a vortex beam. In this case, the spiraling phase contour lines as observed in figure 4.2, lead to an additional wedge prism-like phase ramp at the two slits. Depending on the location of the slits (near or far from the core), this effect significantly changes the fringe period observed on a screen.

The detailed structure of higher order vortices may also be studied with the angular spectrum method to understand the rich structure of near-core fields. Using the result in equation (4.7), the vortex wavefront of the form $\exp(il\theta)$ can again be shown to be a product of two terms:

$$u(r, \theta, z) = \exp(il\theta)\ V_l(r, z), \tag{4.12}$$

where

$$V_l(r, z) = i^l l \exp(il3\pi/2) \int_0^\infty \frac{e^{iz\sqrt{k^2 - 4\pi^2\rho^2}}}{\rho} J_l(2\pi r\rho) d\rho \tag{4.13}$$

In figure 4.3 we summarize the results for the total phase $\Phi(r, \theta, z)$ and the phase lag $V_1(r, z)$ for propagating vortices with charges 1, 2, 3. From figure 4.3 we observe that the phase difference between the vortex beam and the plane wave $(kz - \Phi(r, \theta, z))$ increases with increasing the value of the topological charge. The size of the vortex core region is also seen to increase for higher l values as seen from the field amplitude plot shown in figure 4.3(c).

In summary this chapter used the angular spectrum formalism developed in chapter 3 for exploring the nature of the near-core structure of optical vortices. Contrary to the usual paraxial treatment where helical wavefront phase functions of the form $\exp(il\theta)$ are associated with optical vortices, the angular spectrum picture suggests that there is a rich spiralling phase front structure near the core of optical vortices. The angular spectrum picture inherently incorporates all spatial frequencies and can therefore readily handle the fast varying wavefront field features in the near-core region.

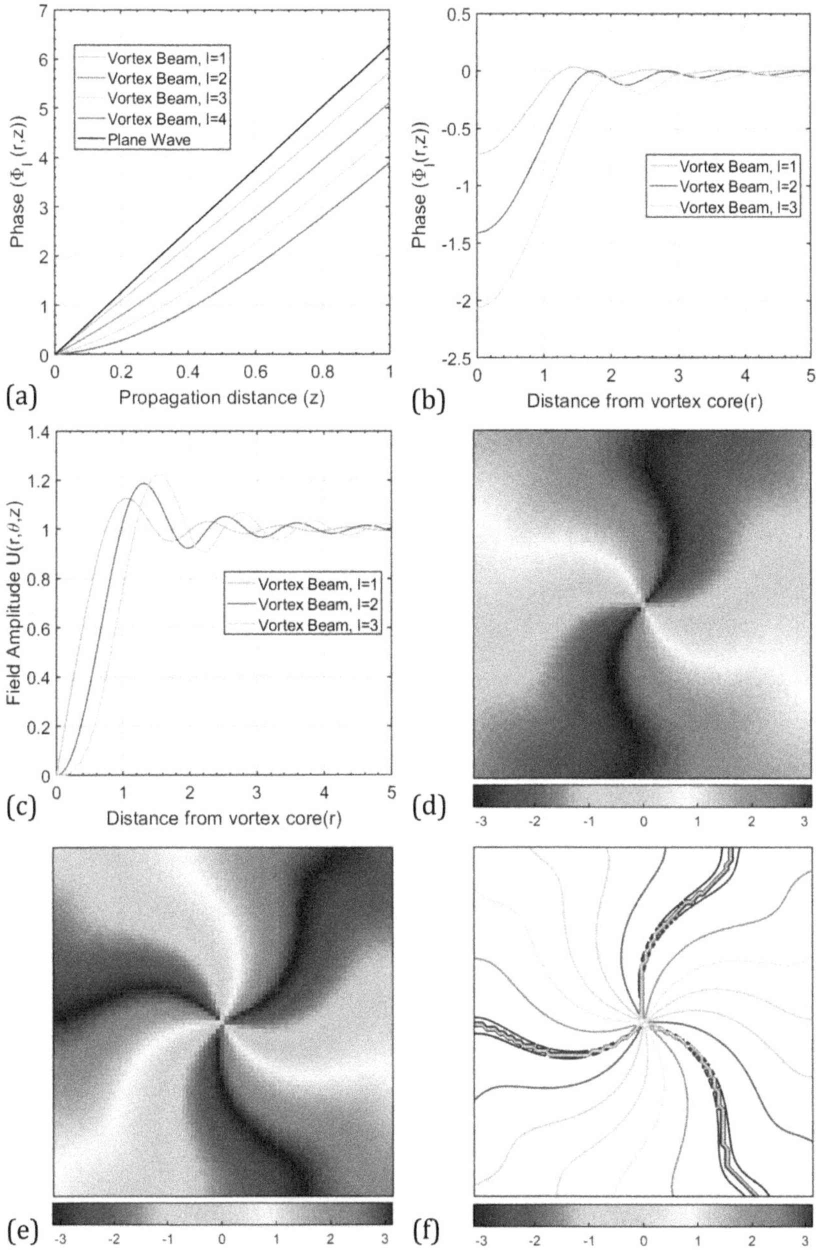

Figure 4.3. The variation of the phase function $\Phi(r, \theta, z)$ for different values of topological charge l is shown. In (a), $\Phi(r, \theta, z)$ (in radians) is plotted as a function of the propagation distance z at $r = 0.1\lambda$. Panel (b) presents $\Phi(r, \theta, z)$ (in radians) as a function of the radial distance r at $z = 1\lambda$. Panel (c) shows the field amplitude $u(r, \theta, z)$ as a function of the radial distance r at $z = 1\lambda$. Panels (d) and (e) display the total phase map $\Phi(r, \theta, z)$ for $l = 2$ and $l = 3$, respectively at a propagation distance of $z = 1\lambda$. Panel (f) illustrates the spiraling phase contour lines for an $l = 3$ vortex. In these plots, the z and r values are given in units of λ. The range of x–y coordinates in (d)–(f) spans from -2λ to 2λ. Adapted with permission from [1], copyright (2016) Optical Society of America.

References

[1] Lochab P, Senthilkumaran P and Khare K 2016 Near-core structure of a propagating optical vortex *J. Opt. Soc. Am.* A **33** 2485–90

[2] Berry M V 2004 Optical vortices evolving from helicoidal integer and fractional phase steps *J. Opt. A: Pure Appl. Opt.* **6** 259–68

[3] Senthilkumaran P and Bahl M 2015 Young's experiment with waves near zeros *Opt. Express* **23** 10968–73

IOP Publishing

Orbital Angular Momentum States of Light (Second Edition)

Propagation through atmospheric turbulence

Kedar Khare, Priyanka Lochab and Paramasivam Senthilkumaran

Chapter 5

Orbital angular momentum states of light

This chapter provides an introduction to the scalar orbital angular momentum states and their generic properties. The azimuthal phase variation is accompanied by the presence of a phase singularity at the center of these beams. The amplitude of the helical wavefront at the on-axis point $r = 0$ vanishes and as explained later a line integral of phase around the zero amplitude point is equal to an integer multiple of (2π). Their wavefront structure is helical or in some cases contains several intertwined helices. The scalar phase singular beams when mixed in orthogonal polarizations can give rise to inhomogeneously polarized vector beams which host variety of polarization singular structures. In the current chapter, we review the properties of scalar orbital angular momentum (OAM) beams.

5.1 Solutions of paraxial wave equation with phase singularities

If the propagation vectors of the optical waves subtend small angles with respect to the propagation axis, then they are called *paraxial waves* [1]. Assuming the nominal propagation direction to be along the z-axis,

$$U(x, y, z) = u(x, y, z)\exp(ikz). \tag{5.1}$$

where $u(x, y, z)$ is a complex quantity and varies slowly with respect to z. The paraxial form of the wave equation can be obtained by substituting equation (5.1) into the Helmholtz equation and neglecting the term containing the second derivative of $u(x, y, z)$ with respect to z. Under the condition that

$$\lambda \left| \frac{\partial^2 u}{\partial z^2} \right| \ll \left| \frac{\partial u}{\partial z} \right|, \tag{5.2}$$

doi:10.1088/978-0-7503-5959-7ch5

one can write the paraxial wave equation in the form:

$$\frac{\partial^2 u}{\partial x^2} + \frac{\partial^2 u}{\partial y^2} + 2ik\frac{\partial u}{\partial z} = 0. \tag{5.3}$$

In cylindrical coordinates, this equation can be written as:

$$\frac{1}{r}\frac{\partial}{\partial r}\left(r\frac{\partial u}{\partial r}\right) + \frac{1}{r^2}\frac{\partial^2 u}{\partial \theta^2} - 2ik\frac{\partial u}{\partial z} = 0 \tag{5.4}$$

With the separation of variables technique, the solutions of this equation are the Laguerre–Gaussian (LG) functions (or modes) [2] given by:

$$\begin{aligned} u(r, \theta, z) = LG(p, l) = &\sqrt{\frac{2^{|l|+1}p!}{\pi(|l| + p)!}}\frac{r^{|l|}\exp(il\theta)}{W_0^{|l|+1}(1 + iz/z_r)^{|l|+1}}\left(\frac{z_r - iz}{z_r + iz}\right)^p \\ &\times \exp\left(\frac{-r^2}{W_0^2(1 + iz/z_r)}\right)L_p^{|l|}\left(\frac{2r^2}{W_0^2(1 + z^2/z_r^2)}\right). \end{aligned} \tag{5.5}$$

Here, W_0 is the $1/e$ width of the Gaussian in the beam waist plane ($z = 0$), $z_r = kW_0^2/2$ is the Rayleigh range of the beam and $|l|$ and p are the azimuthal and radial mode indices associated with the generalized Laguerre polynomial $L_p^{|l|}$,

$$L_p^{|l|}(x) = \frac{e^x x^{-|l|}}{p!}\frac{d^p}{dx^p}[e^{-x}x^{|l|+p}] \tag{5.6}$$

The important quantity to note here is the integer l which gives the topological charge of this beam through the term $\exp(il\theta)$. For $p = 0$, the two lowest order LG modes can be written in $z = 0$ plane as:

$$\mathrm{LG}(0, 0) = \sqrt{\frac{2}{\pi W_0^2}}\exp(-r^2/W_0^2), \tag{5.7}$$

$$\mathrm{LG}(0, 1) = \frac{2r}{\sqrt{\pi}W_0^2}\exp(i\theta)\exp(-r^2/W_0^2). \tag{5.8}$$

The LG(0,1) mode appears as a single annular intensity ring with 2π phase singularity along the beam axis, illustrated in figure 5.1(a). A few other examples of the intensity and phase profile of different LG modes are shown in figures 5.1(b) and (c).

5.2 Orbital angular momentum of LG modes

Light carries both linear momentum and angular momentum. The linear momentum is equal to h/λ per photon while the angular momentum depends on the beam's spatial structure and polarization. The linear momentum density $\boldsymbol{p}$ and the angular momentum density $\boldsymbol{j}$ of a light beam can be obtained from its electric and magnetic field as,

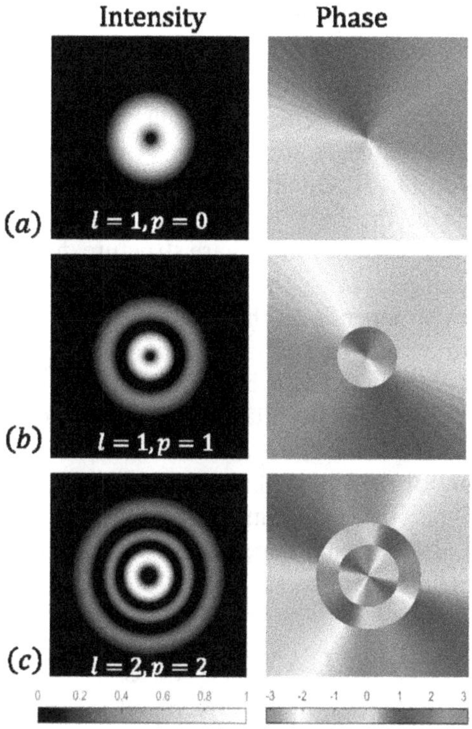

Figure 5.1. The intensity and phase of different LG modes in the $z = 0$ plane is shown. The azimuthal (l) and radial (p) indices take the values (a) $l = 1, p = 0$ (b) $l = 1, p = 1$ and (c) $l = 2, p = 2$, respectively.

$$\boldsymbol{p} = \varepsilon_o(\boldsymbol{E} \times \boldsymbol{B}) \tag{5.9}$$

$$\boldsymbol{j} = \varepsilon_o[\boldsymbol{r} \times (\boldsymbol{E} \times \boldsymbol{B})] = (\boldsymbol{r} \times \boldsymbol{p}) \tag{5.10}$$

In the Lorentz gauge, the field $U(x, y, z)$ in equation (5.1) may be associated with a scalar component of the vector potential $\boldsymbol{A}$. Substituting this field in the expressions for $\boldsymbol{p}$ and $\boldsymbol{j}$ above (equations (5.9) and (5.10)), one may obtain the linear momentum density of a light beam, which is given by [3],

$$\begin{aligned} \boldsymbol{p} &= \frac{\varepsilon_o}{2}(\boldsymbol{E}^* \times \boldsymbol{B} + \boldsymbol{E} \times \boldsymbol{B}^*) \\ &= \frac{i\omega\varepsilon_o}{2}[u^*\nabla u - u\nabla u^*] + \omega k\varepsilon_o\,|U|^2\hat{z} + \frac{\omega\sigma\varepsilon_o}{2}\frac{\partial|u|^2}{\partial r}\hat{\theta} \end{aligned} \tag{5.11}$$

where σ denotes the degree of polarization of light which is equal to ± 1 for circularly polarized light and zero for linearly polarized light. From equations (5.10) and (5.11), we see that the z component of the angular momentum density $\boldsymbol{j}$ depends on the θ component of $\boldsymbol{p}$ as $j_z = rp_\theta$. Therefore, in order to have non-zero j_z value, the light field must have a non-zero linear momentum in the azimuthal direction ($p_\theta \neq 0$).

An infinite plane wave has only transverse components in its electromagnetic field and thus it cannot carry any angular momentum. However, beams of finite size can possess angular momentum in one of two ways. The angular momentum can contain a spin and an orbital contribution. The spin angular momentum (SAM) is associated with the photon spin and manifests itself in the form of circular polarization with a value of $\pm\hbar$ per photon [4, 5]. On the other hand, OAM is associated with the complex spatial profile of the light beam. As discussed in the previous section, optical vortex beams have an azimuthal phase structure given by $\exp(il\theta)$. This gives rise to a circulating component in the k-vector and correspondingly a non-zero OAM [6, 7]. As the OAM depends on the position vector $\boldsymbol{r}$, vortex-free beams with non-zero net OAM can exist. For example, a Gaussian beam would have non-zero OAM if the axis around which OAM is being calculated is different from the beam's axis. However, these beams have un-quantized OAM. The angular momentum for circularly polarized LG modes in the z direction is equal to $\sigma\hbar$ per photon for the SAM and $l\hbar$ per photon for the OAM [8, 9]. The value of SAM is independent of the choice of the $\boldsymbol{r}$ and is therefore called intrinsic. It was shown by Berry [10] that in some cases, the OAM also does not depend on the choice of the beam axis and can be considered to be intrinsic.

5.3 Topological charge of OAM carrying beams

Consider an OAM carrying LG mode with $l \neq 0$ and $p \neq 0$. We observe that at $r = 0$ the field amplitude is zero and as a result the phase is not defined on-axis. Such a zero amplitude point is called a phase singularity if the integral of the phase gradient on a closed path C enclosing the point is not equal to zero [11]:

$$\oint_C \nabla\Theta \cdot d\boldsymbol{r} \neq 0 \tag{5.12}$$

Here Θ is the phase of the LG mode defined in equation (5.5). In the neighborhood of the phase singular point, the gradient of phase therefore has a non-zero curl. For the LG modes, this phase change is an integer multiple of 2π and this integer is commonly known as the *topological charge* [8]. The topological charge denoted by l, is defined as [12–14]:

$$l = \frac{1}{2\pi}\oint_C \nabla\Theta \cdot d\boldsymbol{r} \tag{5.13}$$

The topological charge can take both positive and negative values. The sign of the topological charge describes the handedness of the helical phase. The phase increases in anti-clockwise sense for a positive l vortex. There can also be cases when l is not an integer. Such a fractional charge beam can be expressed as a linear combination of the LG modes. The phase distribution of an LG(0, l) mode of charge l in $z = 0$ plane is given by $\Theta(r, \theta, 0) = l\theta + k_z z = l\theta$ thus giving a phase gradient of $\nabla\Theta = \hat{\theta}/r + k_z\hat{z}$. We see that in any transverse plane the phase gradient has only azimuthal component. This circulating phase gradient is shown in figure 5.2. The phase gradient $\boldsymbol{\nabla}\Theta$ points in the direction of maximum increase of phase and is normal to the phase contour surface.

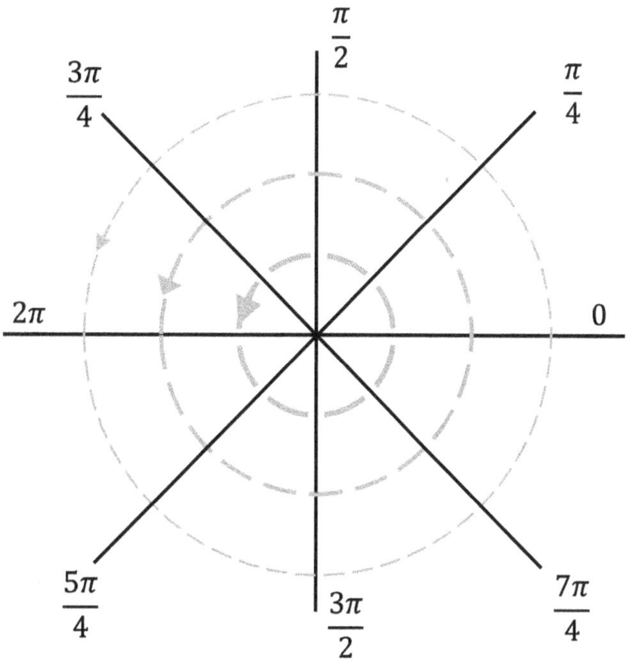

Figure 5.2. The transverse circulating phase gradient is shown (in orange) for a right-handed ($l = 1$) optical vortex. The strength of the phase gradient decreases as one moves radially away from the core. The phase contour lines representing constant phase are also shown (in black).

Besides this, one can also look at the phase contour lines in figure 5.2. These equiphase lines are seen to terminate on the optical vortex in this paraxial propagation model. A peculiarity of the phase front of the OAM beam is the phase structure of form $l\theta$ which makes the propagating wavefront surface helical in nature. There is one major difference in the wavefront of these beams as compared to other non-singular beams. For other non-singular beams, the equiphase surfaces are separated in space from one another, whereas in singular beams there is no concept of distinct separate wavefronts, instead the helical wavefront is joined together to form a common spiralling surface stretching along the propagation direction. For optical vortices with $|l| > 1$, the wavefront structure has $|l|$ intertwined helical surfaces. The phase singularity is present only at a single on-axis point and this manifests as lines of darkness on propagation in three-dimensional space.

5.4 Generation of OAM beams

Optical vortices occur naturally in speckles of coherent laser beams. These can also be observed in the dark regions of three or more multiple beams interference pattern. However, in order to study the general structure and properties of optical vortices, it is first necessary to be able to generate pure OAM modes in a controlled manner in the laboratory. Only then can one venture into harnessing their unique properties for specific applications. Over the last decades, many methods have been developed for generation of OAM beams. These methods can be broadly classified in two categories: (a) methods which try to manipulate the phase of an input beam to generate an azimuthal phase dependence, and (b) modifying laser cavities so that output produced is itself an LG mode. There exist a large number of optical vortex generation methods, however, here we describe only those methods in detail which are relevant to the experiments reported here.

The first type of generation methods involve use of specially designed phase elements, like spiral phase plates, wedges, diffractive elements like fork gratings, spiral zone plates, Dammann vortex gratings, adaptive helical mirrors and spatial light modulators (SLMs) to modify the phase profile of the incoming beam to produce an azimuthal phase dependence. The LG modes generated from these methods are usually not 100% pure and might also contain unwanted higher order LG modes.

5.4.1 Spiral phase plate

Consider a case where a beam whose optical field is given by $u(\boldsymbol{r})$ is incident on a phase plate. For beams with small divergence and phase plates with sufficiently small height, the waves emerging from the plate remain in the paraxial regime. In this case, the operation of the phase plate can be considered to be an operation on only the phase and the transmitted field $u'(\boldsymbol{r})$ is given by:

$$u'(\boldsymbol{r}) = u(\boldsymbol{r})\exp(i\psi) \tag{5.14}$$

where ψ is the phase gained on passing through the plate. Therefore, in order to have an azimuthal phase term, it is natural to use a phase plate with helical shape. The spiral phase plate [15–17] is such an optical element and is one of the simplest methods of introducing an azimuthal phase dependence in the input beam. The spiral phase plate is a transparent plate whose thickness varies proportionally to the azimuthal angle θ around a point in the middle of the plate. The thickness along any radial line remains constant. This type of plate for optical wavelengths was first demonstrated by Beijersbergen *et al* [16] in 1994. Figure 5.3 shows the typical

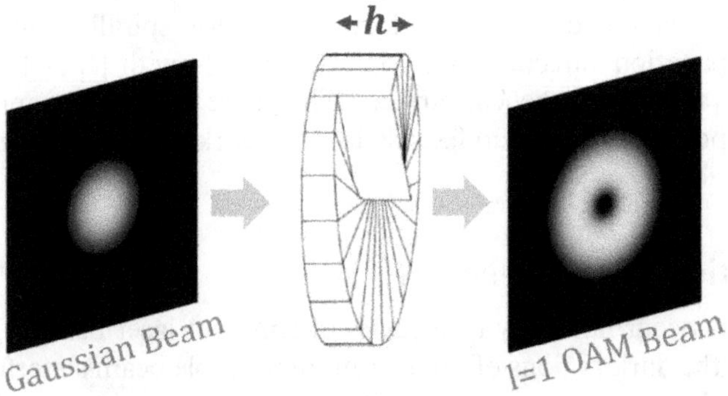

Figure 5.3. Illustration of a spiral phase plate in action, demonstrating the conversion of a Gaussian beam into an OAM state with topological charge $l = 1$. The spiral phase plate imparts a helical phase structure to the beam, resulting in the characteristic twisted wavefront associated with the $l = 1$ OAM mode.

structure of such a phase plate. Let the height difference between the thickest and the thinnest portion of the spiral phase plate be h, then the thickness t of the plate producing a positive charge vortex, can be written as a function of the azimuthal phase angle θ:

$$t(\theta) = \frac{h\theta}{2\pi} \tag{5.15}$$

Assuming the vacuum wavelength of the input beam to be λ and the refractive indices of the spiral phase plate and the background material to be n and n_o respectively, then the optical path length traversed by the beam is given by:

$$\text{Path Length} = \left(h - \frac{h\theta}{2\pi}\right)n_o + \frac{h\theta}{2\pi}n \tag{5.16}$$

Neglecting the overall constant contribution, the phase ψ gained by the beam on passing through the spiral plate is:

$$\psi(\theta) = k \times \text{Path Length} = \frac{h\theta}{\lambda}(n - n_o) \tag{5.17}$$

If the plate can be fashioned in such a way that the factor

$$\frac{h}{\lambda}(n - n_o) = l \tag{5.18}$$

where l is an integer, then, the field $u'(\mathbf{r})$ obtains a helical phase,

$$u'(\mathbf{r}) = u(\mathbf{r})\exp(il\theta) \tag{5.19}$$

This beam on propagation acquires a central intensity null. Therefore, spiral phase plates can be used to convert non-helical beams into helical beams and can also be utilized to change the helicity of any given helical beam. However, the modes generated are not pure modes as only the phase of the incident beam is modified and not the amplitude distribution, for example, when an LG(0,0) beam falls on an $l = 1$ spiral phase plate, the azimuthal index (l) of the LG mode changes to $l = 1$, however, it does not produce a pure LG(0,1) mode and in fact contains contributions from higher order LG modes of the same azimuthal order l but different radial (p) orders. It has been be shown that only 78.5% of the beam is in LG(0,1) mode [16]. The fabrication of a continuous phase ramp is very difficult at optical wavelengths and it is usually approximated as a staircase structure which causes further loss in mode purity. Another limiting factor is the dependence of the spiral phase plate on the input beam wavelength (see equation (5.18)). Recently, new methods like photo polymerization [18–20] and micro-machining techniques [21, 22] have been developed for manufacturing spiral phase plates.

5.4.2 Diffractive optics

This method involves the use of diffractive optical elements (DOEs) to convert coherent non-singular beams to beams containing optical vortices. Two very well-known examples are fork grating [23, 24] and spiral zone plates [25, 26]. The forked design can be implemented as amplitude or phase gratings. In the case of forked grating, multiple orders ($\approx n$) with azimuthal dependence of $\exp(in\theta)$ are obtained with each diffraction order diffracted at a specific angle. This spatial separation of the orders makes it easier to spatially filter and use the desired vortex order. An example of fork grating and spiral zone plate for producing vortices of charges $l = 1$ and $l = 3$ is shown in figure 5.4. These DOE structures may be implemented in the form of computer-generated holograms (CGHs) [27–29]. The forked grating and the spiral zone plates can be generated by interfering a helical vortex beam with a plane and a spherical wave, respectively [30]. The image of the obtained interference pattern can then be simply printed onto a film or mask, thus bypassing the need to physically record them. Just like spiral phase plates, the diffraction orders produced by these diffractive elements are not pure LG modes and contain superposition of many LG states [31]. The CGH transparency mask can, however, be used with multiple wavelengths.

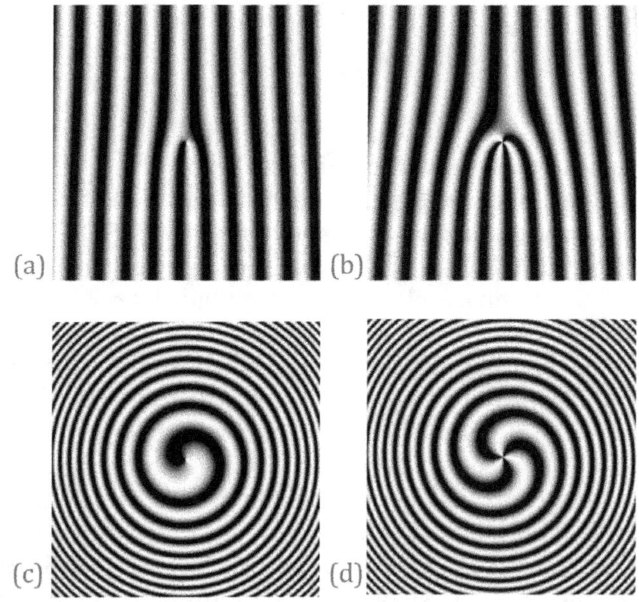

Figure 5.4. Sinusoidal amplitude fork grating and spiral zone plate are shown for topological charges: (a) and (c) for $l = 1$, and (b) and (d) for $l = 3$. Panels (a) and (b) display the fork grating, while panels (c) and (d) illustrate the spiral zone plate.

5.4.3 Spatial light modulators

The CGHs for generating vortex carrying beams can also be implemented on SLMs. SLM is a pixellated device that can modulate the phase, amplitude and polarization of the incident input beam [32–35]. There are various types of SLMs which differ from each other by the manner in which the phase modulation is controlled (electronically, optically or magnetically) [36–41]. The most common SLMs are the electronically controlled phase-only devices [42, 43]. The SLMs work in reflective mode and make use of liquid crystals to introduce the desired phase retardation into the input beam. The refractive index of the liquid crystals can be modified by applying a suitable electric field. The light reflecting from different pixels of the SLM device encounters different refractive indices, thus acquiring the desired phase difference between them. Many techniques have been developed to utilize these phase-only SLMs for carrying out amplitude [44, 45] as well as complex amplitude modulation [46, 47] of the input beam as required in several applications. Therefore, any arbitrary beam can be obtained by displaying the required CGH produced on the SLM. Unlike the traditional DOEs, SLMs offer a dynamic reconfigurable alternative for producing complex fields. Another interesting property of the liquid-crystal SLMs is that they are often sensitive to only one linear polarization. If the input beam is polarized at 45 degrees to the preferred orientation (as may be specified by the SLM manufacturer), then only the polarization component parallel to the SLM axis (assume y-polarization) interacts with the SLM and acquires the displayed phase pattern. The SLM acts like a plane mirror for the other (x-polarized) component of the beam and simply reflects it back without any phase modulation.

5.4.4 Mode converters

This method uses cylindrical lenses to convert a Hermite–Gaussian (HG) mode into a LG mode of the same order [48, 49]. A cylindrical lens changes the shape of the incoming wavefront in only one direction. Therefore, the beam would acquire two different Gouy phases along the x and y directions. By fine-tuning this Gouy phase difference between a pair of HG modes, one can produce a pure LG mode. The second cylindrical lens is used to remove the asymmetry in the beam's curvature. Figure 5.5 shows the conversion of a diagonal HG mode into a LG mode.

5.4.5 Other methods

Optical vortices can also be generated by reflecting them from adaptive helical mirrors [50]. It is also known that three or more beam interference gives rise to dark points where complete destructive interference takes place. These points are actually optical vortices. Therefore, one can generate optical vortex arrays by three beam interference methods [51–53] by suitably fine-tuning the tilt between the interfering beams. Direct generation of OAM modes from a laser cavity has also been demonstrated [54–58].

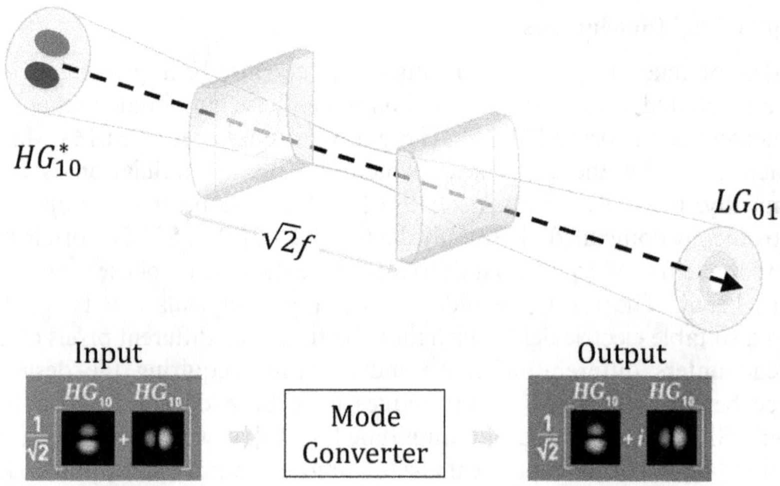

Figure 5.5. The conversion of an HG mode into a LG mode by a pair of cylindrical lenses is shown.

5.5 Detection of phase singularities

Detection of optical vortices is a vast and important research field in itself [13, 14] and some important detection schemes are provided in this section.

5.5.1 Interference-based methods

Since the vortex beams are characterized with an azimuthal phase dependence, interferometry is a natural tool for detecting this phase profile. The vortex beam on interfering with the plane (spherical) beam produces interference fringes with a characteristic fork (spiral) pattern, as shown in figure 5.4. The number of extra arms in the fork (spiral) pattern gives the value of the topological charge of the vortex beam. A Mach–Zender interferometer or lateral shear interferometer can be set up for testing a vortex beam.

5.5.2 Diffraction based methods

Phase also plays an important role in diffraction of beam through apertures. Vortex beams produce unique diffraction patterns on passing through apertures. These diffraction patterns are very distinct from the ones produced by non-singular beams and so can be used to identify a specific OAM state. Diffraction through different types of apertures like single slit [59], double slit [60], diffraction gratings [61], circular [62] and triangular apertures [63, 64], etc has been studied. The vortex beam diffracting through a triangular aperture produces a diffraction pattern with number of lobes equal to the absolute value of the topological charge plus one ($l + 1$) [63]. The helicity of the beam can also be identified by looking at the orientation of the diffraction pattern. Apertures of broken symmetry, i.e. apertures of shape other than square or circular, produce diffraction patterns which give information about the charge of the beam. Recently, many such apertures, like elliptical annular [65], diamond shaped [66] and hexagonal apertures [67] have been studied.

5.5.3 OAM detection using lens aberrations

Lenses which suffer from Seidel aberrations like astigmatism, coma etc are useful for detecting vortex beams [14, 68, 69]. Lenses with circularly non-symmetric aberrations are more likely to break the symmetry of the vortex field and the resultant far-field intensity pattern can thus be used to identify the vortex beam and its charge.

5.5.4 Shack–Hartmann wavefront sensor

A Shack–Hartmann wavefront sensor is a device which consists of a two-dimensional array of lenslets of the same focal length [70]. Each of these lenslets produces a focal spot onto a two-dimensional detector array placed behind them. For an input plane beam, each of the focal spots lies in the center of each lens sub-aperture. However, this focal spot shifts from the center if any tilt or phase variation is present in the input beam. This collection of displacement of the individual focal spots is then used to estimate the wavefront of the input beam. This device has been successfully used as a vortex detector [71, 72]. It is important, however, to note that sufficient sampling of the wavefront by means of lenslet array is required.

5.6 Propagation dynamics of beams embedded with vortices

The propagation dynamics of optical vortices embedded into a host or carrier beam have been studied for linear, non-linear and turbulent media. Consider a host beam, for example a LG(0,0) mode, in which optical vortices have been introduced through one of the methods as discussed earlier. These vortices can be centered on the host beam's axis or they can be distributed off-axis in the beam. In a two-dimensional transverse cross-section of the beam, optical vortices appear as points of null intensity, but when propagated along the z-axis, these null intensity points form lines of darkness. The change in the position of the vortex inside the host beam in different z planes is referred to as the propagation dynamics of the vortex. The trajectories taken by the optical vortices are greatly influenced by their placement in the beam, their neighboring optical vortices and the host beam's phase and intensity profile. Free-space propagation of an array of optical vortices nested in a Gaussian beam was numerically studied by Indebetouw [73] in the paraxial regime. This study showed that when the vortex array consisted of vortices of the same charge, the position of these vortices relative to each other and the host beam do not change on propagation. However, the pattern as a whole expands or contracts together with the host beam. The vortex pattern in the far field is also seen to undergo a rigid rotation by $\pi/2$ in the right (left) handed sense for the array of positively (negatively) charged vortices. In the case of two vortices of opposite charge, their trajectories are seen to depend on the relative value of their initial separation distance $2d$ and the host beam's waist W_0. Generally, the oppositely charged vortices tend to attract each other as they propagate from the waist and pairs of vortices starting sufficiently close to each other ($d < 0.5W_0$) may collide, interfere destructively and then all together vanish from the beam. Oppositely charged vortices which are sufficiently far away in the beam ($d > 0.5W_0$) might survive during propagation to the far field. A similar propagation study was reported in [74], where the dynamical behavior of optical vortices embedded in a plane wave was studied numerically using the angular spectrum method. It was observed that the perturbation caused in the plane wavefront of the beam due to the presence of a single vortex did not influence the propagation dynamics of this vortex. In this case, the vortex continued to propagate perpendicularly to the wavefront. In other words, what this means is that the vortices do not act on themselves and only the presence of nearby optical vortices alter their trajectories. In the same study, an isopolar (same-charge) vortex pair was observed to gyrate around each other during propagation with their centroid propagating in the direction of the wave propagation. In the second case of a bipolar (opposite charge) vortex pair, the vortices were seen to drift laterally away from the direction of propagation. This pair did not gyrate and the separation distance between the vortices remained constant till the point where the vortices annihilated each other. The paper also mentions the sporadic creation and annihilation of bipolar vortex pairs near the focal points of converging wavefronts or when the interference occurs between different waves. Another study by Rozas *et al* [75] found contrasting differences between the propagation dynamics for $r-$ vortices and tanh vortices. The two types of vortices basically differ in the functional form of their core functions. The optical field for a charge l vortex at $z = 0$ plane can be written as: $U(r, \theta, 0) = A(r, z = 0)\exp(il\theta)$ where $A(r, z = 0)$ describes the core or the amplitude function. The r-vortices have amplitude which is linearly varying function of r, for example

$$A(r, z = 0) = \left(\frac{r}{w_v}\right)^{|l|} \tag{5.20}$$

where vortex core size is given by w_v. Similarly, the tanh vortices have localized core functions given by,

$$A(r, z = 0) = \tanh(r/w_v) \tag{5.21}$$

and are used in description of optical vortex soliton which has been observed in non-linear refractive media. If these two types of vortices are embedded into a Gaussian beam of some waist size w_g, then due to the arbitrary small core size of the tanh vortices, their peak intensity value is greater than that for the r-vortices. Drawing on similarities between the hydrodynamic vortices and the optical vortices of paraxial wave fields, the authors of [75] described the important factors which influence the vortex trajectories. They observed that the transverse wave vector at the center of the optical vortex is unaffected by the vortex itself. The trajectory can only be affected by all other sources of phase and intensity gradient, such as diffracting waves and other vortices. This argument is similar to the theory put forward by Roux [74]. They further observed that like the rotation of point vortices in fluids, the tanh vortices when embedded into a Gaussian beam with their cores separated by a distance, orbit each other at rates which depend on the squared vortex separation distance. However, in a self-defocusing non-linear media, very high rotation rates could be achieved. Few other studies on vortex propagation dynamics have also been reported, for example, the cases of vortices with anisotropic phase profiles [76], and vortices in non-linear media [77, 78].

In this chapter we provided the basic description of the scalar OAM states of light and their properties such as topological charge, wavefront structure and propagation dynamics. The methods for generation and detection of OAM states were also discussed. When two distinct OAM states are embedded in orthogonal polarizations, the resultant beam is a vector beam containing a polarization singularity. In a later part of the book, we will present a robust laser beam engineering principle using polarization singular beams. The concept of polarization singularities will be introduced in detail in the next chapter.

5.7 OAM modes as a communication basis

OAM modes as a spatial basis for multiplexing of channels has received much attention in recent years for optical and radio wave communication [79]. OAM basis allows angular momentum as a novel degree of freedom and as a result is a potential candidate to significantly increase information carrying capacity of a communication systems [80–82]. We briefly examine this claim in this section.

In OAM-based multiplexing of information, it is envisioned that OAM basis elements with common optic axis but different charge l can be used to encode different channels. The transmitted wave-function can, for example, be a coherent or incoherent superposition of OAM modes and the information is embedded in time-varying coefficients representing the weights on the individual modes. The receiver may be preferably coaxial with the transmitter. We note that due to the relation:

$$\int_0^{2\pi} d\theta \;\; \exp[i(l_1 - l_2)\theta] = 0, \tag{5.22}$$

for integers $l_1 \neq l_2$, the OAM modes corresponding to the charge l_1 and l_2 are orthogonal to each other on any area centered on $r = 0$ and therefore can be considered at the input end as independent elements for the purpose of information transmission. This relation may suggest use of arbitrarily large number of modes which may be used to encode information at the input end. As the different modes propagate to the receiver, we note that the energy in higher order modes spread over larger rings with on-axis nulls. It is well-known that the spot radius of the OAM mode with indices (l, p) on propagation is given by [83]:

$$r_0(z) = W(z)\sqrt{l + 2p + 1}\,. \tag{5.23}$$

Here $W(z)$ is the spot size expressed in terms of beam waist W_0 as:

$$W(z) = W_0\sqrt{1 + \left(\frac{\lambda z}{\pi W_0^2}\right)^2}\,. \tag{5.24}$$

We therefore observe that the higher order modes denoted by (l, p) have spot sizes that increase proportional to $\sqrt{l + 2p + 1}$. For simplicity let us assume that the on-axis detector radius R_o is small compared to the beam waist W_0 at the detector. For the first two modes the fields are nominally defined as:

$$u_0 = A_0 \exp(-r^2/W_0^2), \tag{5.25}$$

and

$$u_1 = A_1 r \exp(-r^2/W_0^2)\exp(i\theta). \tag{5.26}$$

The power in the zero order mode collected by the detector given by:

$$P_0 = \int_0^{2\pi} \int_0^{R_o} d\theta \; rdr \; |u_0|^2 \tag{5.27}$$

is proportional to $\left[1 - \exp\left(-\frac{2R_o^2}{W_0^2}\right)\right]$. For the first order OAM mode the power P_1 on the same detector is, however, proportional to $\left[1 - \left(1 + \frac{2R_o^2}{W_0^2}\right)\exp\left(-\frac{2R_o^2}{W_0^2}\right)\right]$. In the small detector ($R_o < < W_0$) limit, the ratio P_1/P_0 is given by:

$$\frac{P_1}{P_0} = \left(\frac{R_o}{W_0}\right)^2. \tag{5.28}$$

The on-axis power received from first mode is therefore much smaller than the zero-th mode [84]. The possibility of detecting higher order OAM modes on a finite sized receiver therefore depends on the signal-to-noise performance of the detection mechanism. From purely power considerations it is therefore clear that for a given signal-to-noise, the number of OAM modes that can be effectively detected by an on-axis receiver must be finite and cannot be made as large as we wish.

The question of information carrying capacity of any communication system boils down to examining the spatial degrees of freedom possessed by electromagnetic fields transmitted from and received by a finite sized apertures [85]. For simplicity we assume that the OAM mode transmitter and receiver have apertures of radius R_i and R_o and are separated by a distance z_0 that is much larger than the transmitted and receiver apertures. The receiver for example may be considered in the Fraunhofer zone corresponding to the transmitter. Suppose a waveform $\psi_i(r, \theta)$ in the transmitter plane limited by the aperture of radius R_i. On propagation to the receiver plane in the Fraunhofer plane, the resulting wave-field is described as:

$$\psi_o(\rho, \phi) = C_0 \int\int_{R_i} rdr \; d\theta \; \psi_i(r, \theta) \exp\left[i2\pi r \frac{\rho}{\lambda z_0} \cos(\phi - \theta)\right]. \tag{5.29}$$

In the above equation (r, θ) and (ρ, ϕ) are polar coordinates in the transmitter and receiver planes, respectively, and $C_0 = i \exp(ikz_0)/(\lambda z_0)$. The two functions ψ_i and ψ_o are related by a two-dimensional Fourier transform due to the Fraunhofer approximation used here. Based on the earlier discussion of power received on the detector, we would like to have an input field such that it has maximal energy concentration within the receiver radius R_o. In other words, we would like to maximize the energy concentration ratio:

$$E_r = \frac{\int \int_{R_o} \rho d\rho \, d\phi \, |\psi_o(\rho, \phi)|^2}{\int \int_{R_i} r dr d\theta \, |\psi_i(r, \theta)|^2}. \tag{5.30}$$

The optimal set of orthogonal functions that maximize this ratio are well-known to be the generalized prolate spheroidal functions (GPSFs) [86]. An interesting property of GPSFs is that they are eigenfunctions in the sense that if a GPSF is present in the transmit aperture, it will reproduce itself in the receive aperture with a constant multiplier μ. Further, the number of significant eigenvalues μ is of the order of the space-bandwidth product of the system under consideration. The corresponding GPSFs can retain maximal energy in the receiving aperture. In the finite transmitter and receiver geometry, OAM basis set is not necessarily the most optimal. In fact, in terms of number of modes that may be transmitted, OAM basis may not have any particular advantage over other similar basis sets, for example, the Hermite–Gaussian basis set which is also a solution of the paraxial wave equation [87, 88].

At the receiver end the numerical aperture of the transmitting aperture is given by:

$$NA \approx \frac{R_o}{z}, \tag{5.31}$$

which corresponds to the highest spatial frequency of

$$f_{\max} = \frac{NA}{\lambda}, \tag{5.32}$$

where λ is the illumination wavelength. The basic space-bandwidth product considerations suggest us that the two-dimensional degrees of freedom N_0 that can be supported between such a system are given by:

$$N_0 \approx \pi \frac{R_i^2 R_o^2}{\lambda^2 z^2}. \tag{5.33}$$

This would then be the maximum number of modes that may be supported by such a system irrespective of the basis set used for communication. The above considerations about degrees of freedom are for free space. In the presence of realistic turbulence even over several hundred meters of range, the OAM mode structure may get disturbed beyond recognition. The mode cross-talk in such cases will pose severe limitation for any communication system. We will not discuss this topic any further as it is not a major focus of the book. We will, however, discuss in detail the propagation of OAM states through atmospheric turbulence and present an interesting robust beam engineering principle later in the book that uses OAM states in orthogonal polarizations.

References

[1] Goodman J W 2016 *Introduction to Fourier Optics* 3rd edn (Boston, MA: Roberts)

[2] Dennis M R, O'Holleran K and Padgett M J 2009 Optical vortices and polarization singularities *Prog. Opt. (Ed: E. Wolf)* **53** 293–363

[3] Allen L and Padgett M 2008 Introduction to phase-structured electromagnetic waves *Structured Light and Its Applications* ed D L Andrews (Burlington: Academic) ch 1 1–17 pp

[4] Poynting J H 1909 The wave motion of a revolving shaft, and a suggestion as to the angular momentum in a beam of circularly polarised light *Proc. R. Soc.* **82** 560–7

[5] Beth R A 1936 Mechanical detection and measurement of the angular momentum of light *Phys. Rev.* **50** 115–25 7

[6] Allen L, Padgett M J and Babiker M 1999 The orbital angular momentum of light *Progress in Optics* **vol 39** ed E Wolf (Amsterdam: Elsevier) 291 p

[7] Allen L, Beijersbergen M W, Spreeuw R J C and Woerdman J P 1992 Orbital angular momentum of light and the transformation of Laguerre-Gaussian laser modes *Phys. Rev.* A **45** 8185–9 6

[8] Soskin M S, Gorshkov V N, Vasnetsov M V, Malos J T and Heckenberg N R 1997 Topological charge and angular momentum of light beams carrying optical vortices *Phys. Rev.* A **56** 4064–75 11

[9] Courtial J, Dholakia K, Allen L and Padgett M J 1997 Gaussian beams with very high orbital angular momentum *Opt. Commun.* **144** 210–3

[10] Berry M V 1998 Paraxial beams of spinning light *Int. Conf. on Singular Optics* **vol 3487** ed M S Soskin (Bellingham, WA: International Society for Optics and Photonics, SPIE) 6–11

[11] Soskin M S and Vasnetsov M V 2001 Chapter 4 singular optics *Progress in Optics* **vol 42** ed E Wolf (Amsterdam: Elsevier) 219–74 pp

[12] Nye J F, Berry M V and Frank F C 1974 Dislocations in wave trains *Proc. R. Soc.* 336 165–90

[13] Gbur G 2015 *Singular Optics* (Boca Raton, FL: CRC Press)

[14] Senthilkumaran P 2018 Singularities in physics and engineering Properties, methods and applications IOP Series in Advances in Optics *Photonics and Optoelectronics* (Bristol: IOP Publishing)

[15] Kristensen M, Beijersbergen M W and Woerdman J P 1994 Angular momentum and spin-orbit coupling for microwave photons *Opt. Commun.* **104** 229–33

[16] Beijersbergen M W, Coerwinkel R P C, Kristensen M and Woerdman J P 1994 Helical-wavefront laser beams produced with a spiral phaseplate *Opt. Commun.* **112** 321–7

[17] Turnbull G A, Robertson D A, Smith G M, Allen L and Padgett M J 1996 The generation of free-space Laguerre-Gaussian modes at millimetre-wave frequencies by use of a spiral phaseplate *Opt. Commun.* **127** 183–8

[18] Knöner G, Parkin S, Nieminen T A, Loke V L Y, Heckenberg N R and Rubinsztein-Dunlop H 2007 Integrated optomechanical microelements *Opt. Express* **15** 5521–30 4

[19] Oemrawsingh S S R, van Houwelingen J A W, Eliel E R, Woerdman J P, Verstegen E J K, Kloosterboer J G and 't Hooft G W 2004 Production and characterization of spiral phase plates for optical wavelengths *Appl. Opt.* **43** 688–94 1

[20] Sueda K, Miyaji G, Miyanaga N and Nakatsuka M 2004 Laguerre-Gaussian beam generated with a multilevel spiral phase plate for high intensity laser pulses *Opt. Express* **12** 3548–53 12

[21] Watanabe T, Fujii M, Watanabe Y, Toyama N and Iketaki Y 2004 Generation of a doughnut-shaped beam using a spiral phase plate *Rev. Sci. Instrum* **75** 5131–5

[22] Yu Tsai H, Smith H I and Menon R 2007 Fabrication of spiral-phase diffractive elements using scanning-electron-beam lithography *J. Vac Sci. Technol.* **25** 2068–71

[23] Basistiy I V, Bazhenov V Y, Soskin M S and Vasnetsov M V 1993 Optics of light beams with screw dislocations *Opt. Commun.* **103** 422–8

[24] Bazenhov V Y, Vasnetsov M V and Soskin M S 1990 Laser beams with screw dislocations in their wavefronts *JETP Lett.* **52** 429–31

[25] Sharma M K, Singh R K, Joseph J and Senthilkumaran P 2013 Fourier spectrum analysis of spiral zone plates *Opt. Commun.* **304** 43–8

[26] Vickers J, Burch M, Vyas R and Singh S 2008 Phase and interference properties of optical vortex beams *J. Opt. Soc. Am.* A **25** 823–7 3

[27] Heckenberg N R, McDuff R, Smith C P, Rubinsztein-Dunlop H and Wegener M J 1992 Laser beams with phase singularities *Opt. Quantum Electron.* **24** S951–62 9

[28] Heckenberg N R, McDuff R, Smith C P and White A G 1992 Generation of optical phase singularities by computer-generated holograms *Opt. Lett.* **17** 221–3 2

[29] Carpentier A V, Michinel H, Salgueiro J R and Olivieri D 2008 Making optical vortices with computer-generated holograms *Am. J. Phys.* **76** 916–21

[30] Senthilkumaran P, Masajada J and Sato S 2012 Interferometry with vortices *Int. J. Opt.* **2012** 517591

[31] Sacks Z S, Rozas D and Swartzlander G A 1998 Holographic formation of optical-vortex filaments *J. Opt. Soc. Am.* B **15** 2226–34 8

[32] Fisher A D and Lee J N 1986 The current status of two-dimensional spatial light modulator technology *Proc. SPIE* **0634** 352–71

[33] Efron U 1994 *Spatial Light Modulator Technology: Materials, Devices, and Applications Optical Science and Engineering* (Boca Raton, FL: CRC Press)

[34] Saleh B E and Teich M C 1991 *Fundamentals of Photonics* (New York: Wiley)

[35] Fukushima S, Kurokawa T and Ohno M 1991 Real-time hologram construction and reconstruction using a high-resolution spatial light modulator *Appl. Phys. Lett.* **58** 787–9

[36] Igasaki Y, Li F, Yoshida N, Toyoda H, Inoue T, Mukohzaka N, Kobayashi Y and Hara T 1999 High efficiency electrically-addressable phase-only spatial light modulator *Opt. Rev.* **6** 339–44 6

[37] Moddel G, Johnson K M, Li W, Rice R A, PaganoStauffer L A and Handschy M A 1989 High-speed binary optically addressed spatial light modulator *Appl. Phys. Lett.* **55** 537–9

[38] Clark T W, Offer R F, Franke-Arnold S, Arnold A S and Radwell N 2016 Comparison of beam generation techniques using a phase only spatial light modulator *Opt. Express* **24** 6249–64

[39] Hudson T D and Gregory D A 1991 Optically-addressed spatial light modulators *Opt. Laser Technol.* **23** 297–302

[40] Chung K H, Heo J, Takahashi K, Mito S, Takagi H, Kim J, Lim P B and Inoue M 2008 Characteristics of magneto-photonic crystals based magneto-optic slms for spatial light phase modulators *J. Magnetics Soc. Japan* **32** 114–6

[41] Davis J A, Carcole E and Cottrell D M 1996 Intensity and phase measurements of non-diffracting beams generated with a magneto-optic spatial light modulator *Appl. Opt.* **35** 593–8 2

[42] Collings N, Davey T, Christmas J, Chu D and Crossland B 2011 The applications and technology of phase-only liquid crystal on silicon devices *J. Display Technol.* **7** 112–9 3

[43] Vaupotic N, Pavlin J and Cepic M 2013 Liquid crystals: a new topic in physics for undergraduates *Eur. J. Phys.* **34** 745–61
[44] Davis J A, Cottrell D M, Campos J, Yzuel M J and Moreno I 1999 Encoding amplitude information onto phase-only filters *Appl. Opt.* **38** 5004–13
[45] Márquez A, Iemmi C, Escalera J C, Campos J, Ledesma S, Davis J A and Yzuel M J 2001 Amplitude apodizers encoded onto fresnel lenses implemented on a phase-only spatial light modulator *Appl. Opt.* **40** 2316–22 5
[46] Arrizón V, Ruiz U, Carrada R and González L A 2007 Pixelated phase computer holograms for the accurate encoding of scalar complex fields *J. Opt. Soc. Am.* A **24** 3500–7
[47] de Bougrenet J L, Tocnaye de la and Dupont L 1997 Complex amplitude modulation by use of liquid-crystal spatial light modulators *Appl. Opt.* **36** 1730–41
[48] Beijersbergen M W, Allen L, van der Veen H E L O and Woerdman J P 1993 Astigmatic laser mode converters and transfer of orbital angular momentum *Opt. Commun.* **96** 123–32
[49] Padgett M, Arlt J, Simpson N and Allen L 1996 An experiment to observe the intensity and phase structure of Laguerre-Gaussian laser modes *Am. J. Phys.* **64** 77–82
[50] Ghai D P, Senthilkumaran P and Sirohi R S 2008 Adaptive helical mirror for generation of optical phase singularity *Appl. Opt.* **47** 1378–83 4
[51] Masajada J and Dubik B 2001 Optical vortex generation by three plane wave interference *Opt. Commun.* **198** 21–7
[52] Vyas S and Senthilkumaran P 2007 Interferometric optical vortex array generator *Appl. Opt.* **46** 2893–8 5
[53] Boguslawski M, Rose P and Denz C 2011 Increasing the structural variety of discrete nondiffracting wave fields *Phys. Rev.* A **84** 013832 6
[54] Tamm C 1988 Frequency locking of two transverse optical modes of a laser *Phys. Rev.* A **38** 5960–3 12
[55] Tamm C and Weiss C O 1990 Bistability and optical switching of spatial patterns in a laser *J. Opt. Soc. Am.* B **7** 1034–8 6
[56] Rigrod W W 1963 Isolation of axi-symmetrical optical resonator modes *Appl. Phys. Lett.* **2** 51–3
[57] Ito A, Kozawa Y and Sato S 2010 Generation of hollow scalar and vector beams using a spot-defect mirror *J. Opt. Soc. Am.* A **27** 2072–7 9
[58] Kano K, Kozawa Y and Sato S 2012 Generation of a purely single transverse mode vortex beam from a He-Ne laser cavity with a spot-defect mirror *Int. J. Opt.* **2012** 359141
[59] Ghai D P, Senthilkumaran P and Sirohi R S 2009 Single-slit diffraction of an optical beam with phase singularity *Opt. Lasers Eng.* **47** 123–6
[60] Sztul H I and Alfano R R 2006 Double-slit interference with Laguerre-Gaussian beams *Opt. Lett.* **31** 999–1001 4
[61] Moreno I, Davis J A, Melvin B, Pascoguin L, Mitry M J and Cottrell D M 2009 Vortex sensing diffraction gratings *Opt. Lett.* **34** 2927–9 10
[62] Ambuj A, Vyas R and Singh S 2014 Diffraction of orbital angular momentum carrying optical beams by a circular aperture *Opt. Lett.* **39** 5475–8 10
[63] Hickmann J M, Fonseca E J S, Soares W C and Chávez-Cerda S 2010 Unveiling a truncated optical lattice associated with a triangular aperture using light's orbital angular momentum *Phys. Rev. Lett.* **105** 053904 7
[64] Liu Y, Tao H, Pu J and Lü B 2011 Detecting the topological charge of vortex beams using an annular triangle aperture *Opt. Laser Technol.* **43** 1233–6
[65] Tao H, Liu Y, Chen Z and Pu J 2012 Measuring the topological charge of vortex beams by using an annular ellipse aperture *Appl. Phys.* B **106** 927–32 3

[66] Liu Y, Sun S, Pu J and Lü B 2013 Propagation of an optical vortex beam through a diamond-shaped aperture *Opt. Laser Technol.* **45** 473–9
[67] Liu Y and Pu J 2011 Measuring the orbital angular momentum of elliptical vortex beams by using a slit hexagon aperture *Opt. Commun.* **284** 2424–9
[68] Singh R K, Senthilkumaran P and Singh K 2007 The effect of astigmatism on the diffraction of a vortex carrying beam with a Gaussian background *J. Opt. A: Pure Appl. Opt.* **9** 543–54 5
[69] Kotlyar V V, Kovalev A A and Porfirev A P 2017 Astigmatic transforms of an optical vortex for measurement of its topological charge *Appl. Opt.* **56** 4095–104 5
[70] Platt B C and Shack R 2001 History and principles of Shack-Hartmann wavefront sensing *J. Refract. Surg.* **17** S573–7
[71] Chen M, Roux F S and Olivier J C 2007 Detection of phase singularities with a Shack-Hartmann wavefront sensor *J. Opt. Soc. Am.* A **24** 1994–2002
[72] Murphy K, Burke D, Devaney N and Dainty J C 2010 Experimental detection of optical vortices with a Shack-Hartmann wavefront sensor *Opt. Express* **18** 15448–60 7
[73] Indebetouw G 1993 Optical vortices and their propagation *J. Mod. Opt* **40** 73–87
[74] Roux F S 1995 Dynamical behavior of optical vortices *J. Opt. Soc. Am.* B **12** 1215–21 7
[75] Rozas D, Law C T and Swartzlander G A 1997 Propagation dynamics of optical vortices *J. Opt. Soc. Am.* B **14** 3054–65 11
[76] Kim G H, Lee H J, Kim J U and Suk H 2003 Propagation dynamics of optical vortices with anisotropic phase profiles *J. Opt. Soc. Am.* B **20** 351–9 2
[77] Kivshar Y S and Luther-Davies B 1998 Dark optical solitons: physics and applications *Phys. Rep.* **298** 81
[78] Swartzlander G A 2001 *Optical Vortex Solitons* Springer Series in Optical Sciences vol 82 *ed S Trillo W (Berlin: Springer)*
[79] Willner A E *et al* 2015 Optical communications using orbital angular momentum beams *Adv. Opt. Photon.* **7** 66–106
[80] Tamburini F, Mari E, Sponselli A, Thidé B, Bianchini A and Romanato F 2012 Encoding many channels on the same frequency through radio vorticity: first experimental test *New J. Phys.* **14** 033001
[81] Yan Y *et al* 2014 High-capacity millimetre-wave communications with orbital angular momentum multiplexing *Nat. Commun.* **5** 1–9
[82] Edfors O and Johansson A J 2011 Is orbital angular momentum (oam) based radio communication an unexploited area? *IEEE Trans. Antennas Propag.* **60** 1126–31
[83] Phillips R L and Andrews L C 1983 Spot size and divergence for Laguerre–Gaussian beams of any order *Appl. Opt.* **22** 643–4
[84] Andersson M, Berglind E and Björk G 2015 Orbital angular momentum modes do not increase the channel capacity in communication links *New J. Phys.* **17** 043040
[85] Chen M, Dholakia K and Mazilu M 2016 Is there an optimal basis to maximise optical information transfer? *Sci. Rep.* **6** 22821
[86] Slepian D 1964 Prolate spheroidal wave functions, Fourier analysis and uncertainty–IV: extensions to many dimensions; generalized prolate spheroidal functions *Bell Syst. Tech. J.* **43** 3009–57
[87] Zhao N, Li X, Li G and Kahn J M 2015 Capacity limits of spatially multiplexed free-space communication *Nat. Photon.* **9** 822
[88] Gaffoglio R, Cagliero A, Vecchi G and Andriulli F P 2017 Vortex waves and channel capacity: hopes and reality *IEEE Access* **6** 19814–22

IOP Publishing

Orbital Angular Momentum States of Light (Second Edition)
Propagation through atmospheric turbulence
Kedar Khare, Priyanka Lochab and Paramasivam Senthilkumaran

Chapter 6

Introduction to polarization singularities

Polarization singularities form an important class of beam structures that can be generated using the orbital angular momentum (OAM) states. As discussed in the previous chapter, the OAM in scalar light beams is generally attributed to phase singularities in the beam although beams with non-zero OAM that are non-singular also exist. When OAM states with different topological charge are embedded in orthogonal polarizations, the resultant beams have polarization singularities as we will discuss in detail in this chapter. In comparison to the subject of phase singularities, the complications arising due to additional parameters of polarization of light, make the subject of polarization singularities relatively hard to follow and as a result the potential of polarization singularities has not yet been explored extensively. This chapter is aimed at providing an in-depth understanding on polarization singularities so that readers will appreciate this fascinating field. In the later part of the book, we will introduce a robust laser beam engineering principle using polarization singular beams. This chapter will therefore provide the essential background for this purpose.

6.1 Polarization state of light beams

Consider a coherent monochromatic plane wave traveling along the z direction. For such a wave the electric field oscillations are restricted to the xy plane. As the beam propagates, the changes occurring to tip of the electric field vector describe the state of polarization (SOP) of the beam [1]. If the electric field oscillations are such that the tip of the electric field vector draws a circle, ellipse or a line in a projected xy plane perpendicular to the beam propagation direction, the beam is said to be circularly, elliptically or linearly polarized, respectively. For paraxial beams, we are mainly concerned with transverse fields that are described by two orthogonal polarizations in the xy plane. The amplitude and phase of the component oscillations of a given SOP during orthogonal decomposition are crucial for determining the resultant polarization state. An equivalent description of SOP is

doi:10.1088/978-0-7503-5959-7ch6

in terms of Stokes parameters that do not deal with the component amplitudes and phases, but instead use directly measurable quantities as we describe next.

6.1.1 Stokes parameters

The SOP of light can be described using Stokes parameters [2, 3] that are expressed using intensity measurements as follows.

$$\begin{aligned} S_0 &= I_x + I_y, \\ S_1 &= I_x - I_y, \\ S_2 &= I_{45^\circ} - I_{-45^\circ}, \\ S_3 &= I_{RCP} - I_{LCP}, \end{aligned} \tag{6.1}$$

where I_x, I_y, $I_{(+45^\circ)}$, $I_{(-45^\circ)}$, $I_{(LCP)}$ and $I_{(RCP)}$ are the component intensities when a given SOP is decomposed into linear states oriented along x, y, $(+45^\circ)$, (-45°) and circular states—left circular polarization (LCP) and right circular polarization (RCP), respectively. Stokes parameters are related as:

$$S_0^2 \geqslant S_1^2 + S_2^2 + S_3^2, \tag{6.2}$$

with the equality holding for fully polarized light. In this chapter we will describe polarization singularities in fully polarized light. For a linearly polarized light the Stokes parameters are $S_3 = 0$ and $S_1^2 + S_2^2 = S_0^2$; for circularly polarized light both $S_1 = 0$ and $S_2 = 0$ and $S_3 = \pm S_0$. The polarization state is considered inhomogenous if the Stokes parameters are functions of the transverse (x, y) position coordinates. As we will discuss later, such inhomogeneous polarization state occurs when the two orthogonal (and temporally coherent) polarization components of a light beam have distinct spatially varying amplitude-phase structure.

6.1.2 Azimuth and ellipticity

Apart from Stokes parameters the SOP can be described by two other parameters—ellipticity χ and azimuth γ. For a fully polarized light ellipticity χ is defined by the relation

$$\tan \chi = \pm \frac{b}{a}, \tag{6.3}$$

where a and b are the major and minor axis of the polarization ellipse, the positive sign is for right-handed ellipse and the negative sign is for the left-handed ellipse. The quantity χ varies between $-\pi/4$ and $+\pi/4$. The azimuth γ is the orientation angle of the major axis of the ellipse with respect to the reference direction (usually x). The azimuth varies between 0 and π. In figure 6.1, various polarization ellipses with different azimuth or/and ellipticity are shown for illustration.

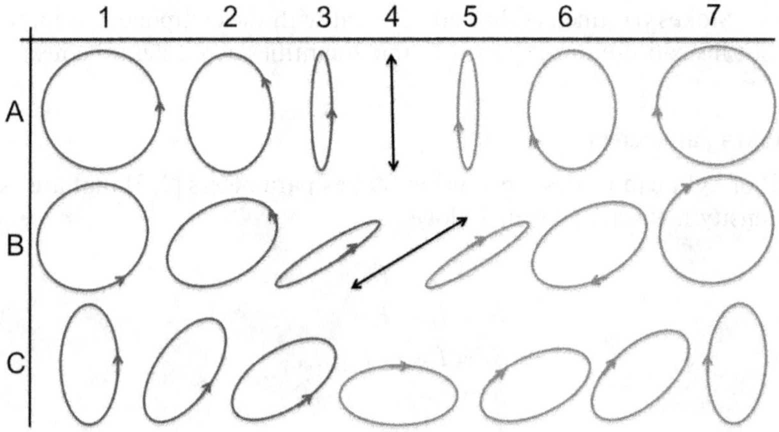

Figure 6.1. Illustration of polarization states with different azimuth and ellipticity. In row A, SOPs starting from right-circularly polarized to left circularly polarized are shown corresponding to ellipticity of $+\frac{\pi}{4}$ to $-\frac{\pi}{4}$. Row B shows states having the same azimuth but varying ellipticity. All the figures in row C show states with the same ellipticity but different azimuth.

6.1.3 Poincare sphere

Using the Stokes parameters, and from relation equation (6.2), it is natural to construct a sphere of radius S_0 in which S_1, S_2, S_3 form three orthogonal axes. This sphere is called the Poincaré sphere and every point on the surface of the sphere represents a particular state of fully polarized light (figure 6.2). Normalized Stokes parameters can be obtained by dividing each of the Stokes parameters by S_0. In the normalized coordinates the Poincaré sphere has unit radius. The North and South poles of the Poincaré sphere represent right and left circularly polarized light, equatorial points represent linearly polarized light with different azimuths and rest of the points represent elliptically polarized light. Points in the northern hemisphere are right-handed polarization states while points in the southern hemisphere are left-handed. We can also use two variables, namely the longitude and latitude [4] to represent any point on the surface of the Poincaré sphere. They are related to Stokes parameters as longitude

$$2\gamma = \arctan\left(\frac{S_2}{S_1}\right) \tag{6.4}$$

and latitude

$$2\chi = \arcsin\left(\frac{S_3}{S_0}\right). \tag{6.5}$$

While homogeneously polarized light can be represented by a point on the Poincaré sphere, it may be noted that inhomogeneously polarized light is represented by a collection of points or regions on the Poincaré sphere.

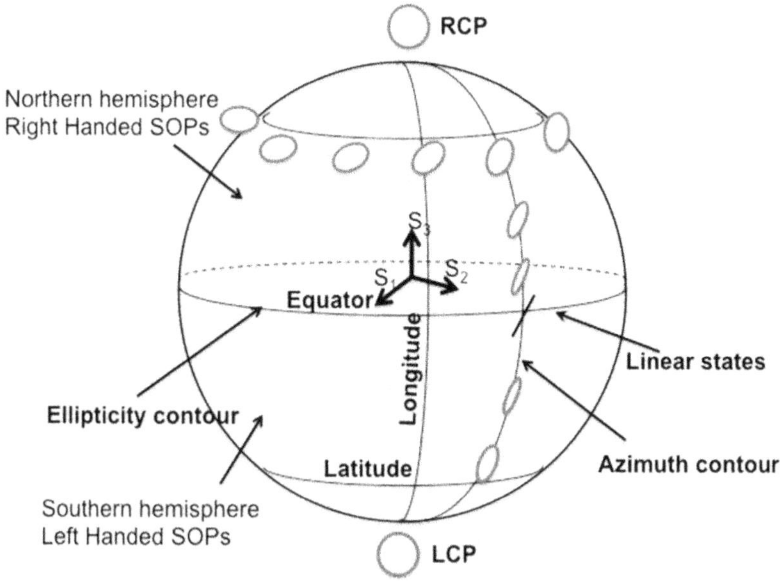

Figure 6.2. Poincaré sphere. North and South poles represent right and left circularly polarized light. Equatorial points represent linear polarization states and the rest of the points on the Poincaré sphere represent elliptically polarized light. The longitudes are azimuth contours and latitudes are ellipticity contours.

6.2 Decomposition of a general SOP

Light in any SOP can be decomposed into two linearly polarized orthogonal states. In other words, any SOP can be represented as a superposition of two linearly polarized states. Consider the following superposition

$$\mathbf{U} = a_x \exp(i\delta_x)\, \hat{x} + a_y \exp(i\delta_y)\, \hat{y} \tag{6.6}$$

Here a_x, a_y are the component amplitudes of the simple harmonic oscillations (for a coherent monochromatic light) occurring in xz and yz planes and δ_x, δ_y are the component phases, respectively. The SOP is circular when $a_x = a_y$ and $\delta_y - \delta_x = (2n + 1)\pi/2$ for an integer n. The SOP is linear when $\delta_y - \delta_x = n\pi$ irrespective of the values of a_x and a_y. The SOP is elliptical in general when linear and circular polarization conditions are not met. Decomposition of SOP in terms of linear basis is explained in most common text books on polarization [1]. It is clear that SOP may also be represented in terms of any other orthogonal basis such as the circular basis [5, 6]. In this case the two component oscillations are clockwise and counter-clockwise rotating circular polarized light oscillations with amplitudes a_R and a_L, respectively. The superposition state here is given by

$$\mathbf{U} = a_R e^{i\delta_R} \hat{e}_R + a_L e^{i\delta_L} \hat{e}_L. \tag{6.7}$$

where δ_R and δ_L are the phases of the component oscillations. In the circular decomposition, the linear states occur when the component oscillations have the same amplitude, i.e., $a_R = a_L$ and elliptical states occur when the component oscillations are such that $a_R \neq a_L$. The orientation of the plane of polarization or the orientation of the major axis of the ellipse is decided by the phase difference between two circular components, $\delta_R - \delta_L$. The following table 6.1 describes the difference between two types of decompositions as presented here. It is important to note that the component amplitude decides the azimuth in linear decomposition, whereas the component amplitudes decide the ellipticity in the case of the circular decomposition. Likewise, the component phases control the ellipticity and the azimuth in the two types of decompositions. The case of circular polarization basis is depicted in figure 6.3 for illustration. The plots along each row in this figure are drawn with the azimuth kept constant. On the other hand, the plots along the column direction are drawn with the ellipticity kept constant. The component amplitudes shown by red color vectors rotate in counter-clockwise direction while the component amplitudes in blue color rotate in clockwise direction. The sense of rotation of component

Table 6.1. Comparison of decomposition using two different basis sets.

S. no.	Decomposition	Component amplitudes	Component phases
1	Linear decomposition	Azimuth control	Ellipticity control
2	Circular decomposition	Ellipticity control	Azimuth control

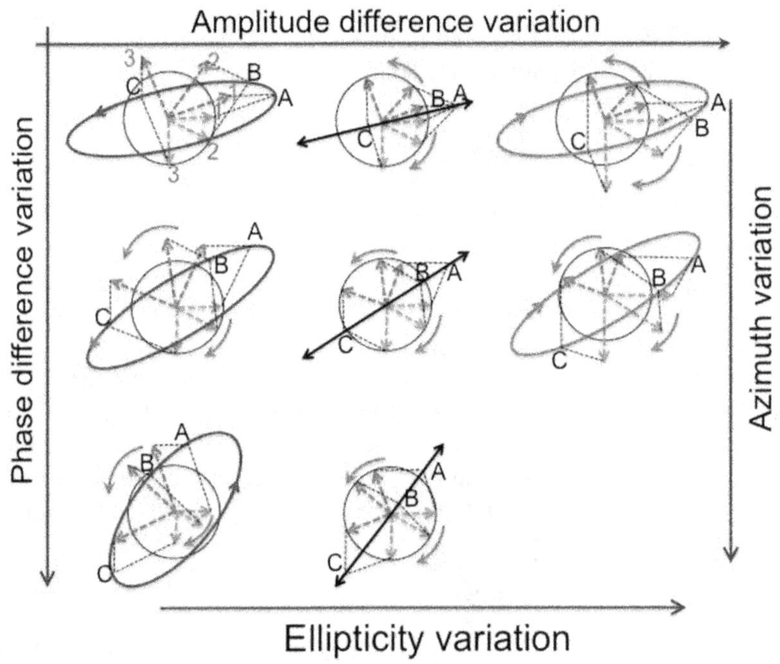

Figure 6.3. Azimuth control and ellipticity control by component phase and amplitude variation, respectively, in circular basis.

vectors is shown by green circulating arrows. As time progresses, the orientations of the circular basis components change in anti-clockwise and clockwise directions for red and blue colored vectors, respectively. At each time instance addition of vectors numbered as red 1 and blue 1 gives the resultant vector denoted by A. Similarly, vectors red 2 and blue 2 are added to give resultant vector B at a later time. Vector C can be constructed in similar fashion with vectors numbered 3. This way the tip of the vector draws an anti-clockwise rotating ellipse and hence the resultant SOP is a right-handed elliptical state. In all the figures the resultant vectors rotate from A to C with increasing time. The component amplitudes are depicted by vectors of different lengths and phase difference between the component vectors are shown by different angular positions the vectors (red 1 and blue 1) take initially.

6.2.1 Helicity and spin

Helicity and spin are different [7, 8] and are not to be treated as synonymous to each other. Photons have integer spin (spin ± 1) with spin angular momentum (SAM) of $\hbar$ for right circular polarized state and $-\hbar$ for left circular polarized state. In the circular basis decomposition equation (6.7), we have seen that any polarization state of light is shown as a superposition of right and left circularly polarized components (superposition of positive and negative spin states). The component amplitudes decide the helicity (or handedness) of the superposition state. If the RCP, component is larger than the LCP component ($a_R > a_L$), then the resulting elliptical polarization state is said to right-handed ellipse state and so on. For linear polarization both the RCP and LCP components are equal ($a_R = a_L$) and therefore there is no handedness associated with the linearly polarized states. If $a_R = 0$, the light is left circularly polarized. Therefore, a left elliptically polarized light has negative helicity (handedness) but has both positive and negative spin components.

6.2.2 Homogeneous and inhomogeneous polarization distributions

Homogeneous polarization refers to uniform polarization distributions in the transverse cross-section of the beam. For example, a circularly polarized beam has homogeneous polarization distribution such that at all points in the beam cross-section the SOP is the same. Similarly, a linearly polarized or an elliptically polarized light refers to uniform polarization across the beam. If a homogeneously polarized beam is passed through a polarizer, then there is a uniform reduction in the intensity of the beam, but there is no change in its spatial amplitude structure. The beams for which there is spatial variation of SOP across the beam cross-section are said to have inhomogeneous polarization and are also referred to as vector beams. Figure 6.4 shows the SOP distribution in a homogenous and inhomogeneous ellipse field for illustration. Beams with slowly varying polarization distributions [9–11] have attracted interest in recent years. For such beams, each of the parameters used to define the SOP is also smoothly varying. For example, the ellipticity can be spatially varying and the ellipticity distribution can have features such as extrema, saddles and so on. Since there are number of parameters such as four Stokes parameters, ellipticity, azimuth, handedness or helicity associated with polarization

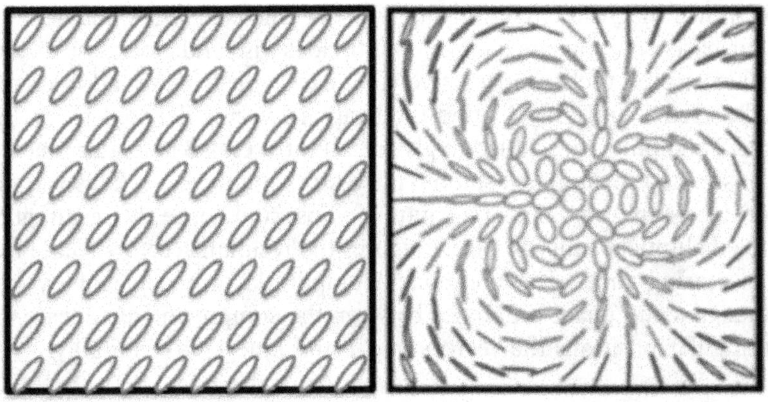

Figure 6.4. Homogeneously polarized beam (elliptically polarized light) and Inhomogeneously polarized ellipse field.

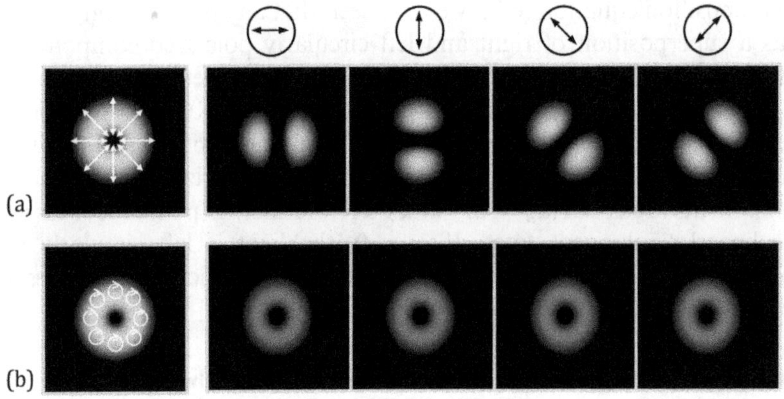

Figure 6.5. The effect of the rotation of polarizer pass axis on scalar and vector modes intensity is illustrated. Here (a) represents an input beam with radial polarization (inhomogeneous polarization), while (b) represents a right-circularly polarized (homogeneous polarization) scalar beam.

of light, one can realize the occurrence of diverse topological features and possibilities in the study of inhomogeneous polarization. This makes the field more interesting and rich with increased level of complexity. When the SOP varies point by point, it is normally illustrated by drawing polarization ellipses at equal spatial intervals (samples). If an inhomogeneously polarized beam is passed through a polarizer, then depending on the angle of the polarizer, one obtains different spatial structure of the beam [12]. The effect of the rotation of polarizer on homogeneous and inhomogeneous polarization distribution is illustrated in figure 6.5. We observe that while both the initial intensity profiles have a similar donut-like structure, the effect of linear polarizer on the two states is completely different.

6.3 Singularities in optical fields

A point within an optical field is called a singular point if a physical parameter associated with the field is not defined or not-well behaved (e.g. goes to infinity) [13]. In the neighborhood of this point the field gradients are typically large in magnitude. In electromagnetic fields, at a phase singularity the phase is not defined and at a polarization singularity the polarization azimuth is not defined. For an elliptically polarized light, azimuth refers to the angle that the major axis of the ellipse makes with respect to a reference direction (say x). For a linearly polarized light, azimuth refers to the angle of orientation of the vibration plane with respect to the reference orientation. A phase singularity is accompanied with a neighborhood in which all phase values ranging from 0 to $2m\pi$ are present for integer m. Similarly, in a polarization singularity the neighborhood polarization distribution is such that the polarization azimuth has all possible orientations. The phase gradient in a phase singularity and azimuth gradient in a polarization singularity circulate around the respective singularities. This also implies that in a phase singularity the phase contours converge at the singular point and in a polarization singularity the azimuth contours converge at the singular point.

6.3.1 Phase singularities

Optical phase singularity was already discussed in chapter 5 in the context of OAM states and is a point-phase defect. The phase singularity is also referred to as optical vortex. The complex amplitude of a phase singular beam is nominally given by:

$$U(r,\,\theta) = f(r)\exp(im\theta). \tag{6.8}$$

where $f(r)$ is the r-dependent part. The index m is called topological charge of the vortex, defined by:

$$m = \frac{1}{2\pi}\oint \nabla\theta \cdot d\mathbf{l} \tag{6.9}$$

It can take positive and negative integer values depending on handedness of the helical wavefront. The phase distribution is given by azimuthally varying function $m\theta$, where θ is the angle in polar coordinate. Laguerre–Gaussian (LG) beam with non-zero azimuthal index is an example of a phase singular beam. At $r = 0$ the amplitude of the LG beam is zero and thus the phase is undefined. Another commonly used example, is an r-vortex in which the complex amplitude is given by $U = (x \pm iy)^m$ where the amplitude distribution is given by $\left(\sqrt{x^2 + y^2}\right)^m$ and the phase distribution is given by $\arctan(y/x)$. The r-vortex can be used to understand many of the properties of phase singularity. The transverse phase gradient [14, 15] for this vortex is given by $\nabla\theta = (m/r)\hat{\theta}$. The phase singular point is characterized by circulating phase gradient and near the vortex core large phase gradient exists [16–18]. As seen in the previous chapter, the phase singular beams carry OAM [19, 20]. This can be seen by analogy with quantum mechanics [21, 22] in which the

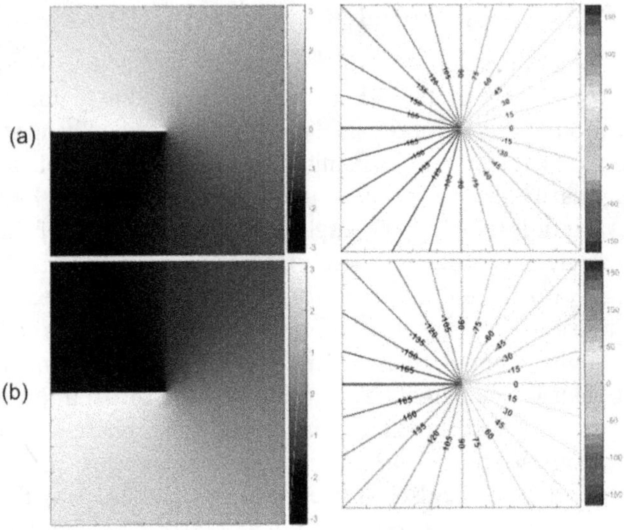

Figure 6.6. Phase distributions and phase contours for a vortex of charge (a) +1 and (b) −1 are shown. The numbers in the right panel figures depict the phase values in degrees, illustrating the spiraling phase structure of vortices.

momentum operator $L_z = -i\hbar\partial/\partial\theta$ acting on the wave function of the form $f(r)\exp(im\theta)$ gives rise to OAM of $m\hbar$. Here the charge m of the vortex decides the angular momentum state. The phase distribution for a positive and negative unit charge vortex and the corresponding phase contours are shown in figure 6.6. The phase contours are radial from the vortex point. The phase gradient is normal to the phase contours and is therefore oriented in the azimuthal direction. The sense of circulation depends on which way the phase increases in the circulation. The amplitude profiles and distributions for an r-vortex (equation (5.20)), tanh-vortex (equation (5.21)) and an LG beam are shown in figure 6.7.

6.3.2 Polarization singularities

The polarization structures of inhomogeneously polarized beams can be very complicated. Just like phase singularities of scalar fields, the inhomogeneously polarized field can also have its own singularities associated with polarization parameters like ellipticity or azimuth [12, 13, 23–25]. These are points in the vector field where some aspect of the beam's polarization is undefined. All vector beams which contain polarization singularities have spatially inhomogeneous polarization distribution. However, the reverse is not necessarily true, i.e., not all inhomogeneously polarized beams contain polarization singularities. The polarization singularities can be divided into two types—(a) elliptic-point and (b) vector-point singularity depending on the SOP of the host beam. There also exits a third type of polarization singularity known as the Stokes point (Σ-point) singularity [26]. At a Σ-point, all three normalized Stokes parameters become undefined. Hence, at these

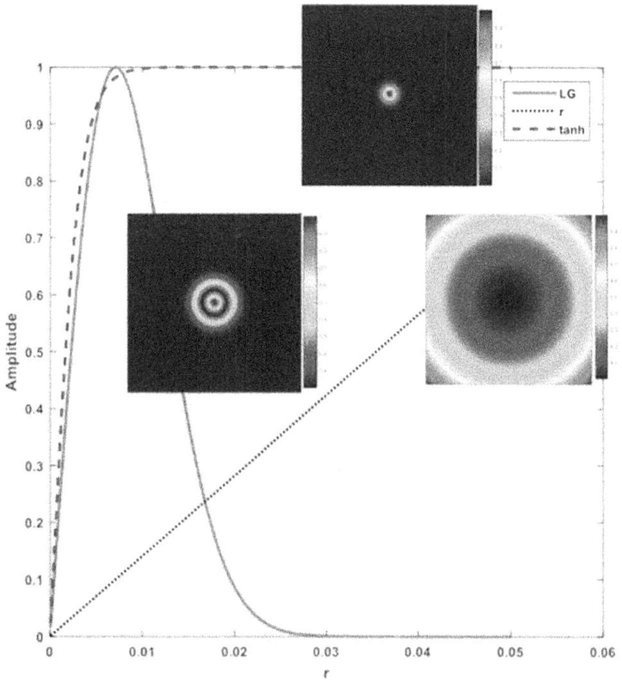

Figure 6.7. Three distinct types of amplitude profiles and their corresponding amplitude distributions for a unit charge vortex are illustrated.

points, the SOP itself is undefined. As of now very little is known about Σ-point singularities. In the present chapter, we will only concentrate on the elliptic-point and vector-point polarization singularities. Elliptic-point singularities are present in inhomogeneous elliptically polarized fields [24, 25]. At these points, some property of the polarization ellipse becomes undefined. The planar elliptic-point singularities are further characterized into two major categories—the L-lines and the C-points. The L-lines category is a continuous line of linear polarization embedded in a field of ellipses. The handedness of the ellipse (right or left) becomes undefined at the L-line. Hence, the L-line usually separates the regions of different handedness in a paraxial elliptically polarized field. The C-points are the points of circular polarization at which the orientation of the major (or minor) axis of the polarization ellipse is undefined. The polarization ellipses surrounding the C-point are usually arranged in three specific arrangements, known as the lemon, monstar and star. Different numbers of polarization lines terminate on these three structures. Only one line terminates on a lemon, three on a star and infinitely many with three straight on a monstar [27]. The elliptical polarization fields for lemon, star and monstar are shown in figure 6.8. In order to explain the singularity index I_c associated with the C-point singularities we show in figure 6.9 how the rotation of the major axis of the elliptical polarization changes around the singularity for positions 1 to 7. The sense in which the major axis of the ellipse rotates is seen to be opposite for the lemon and star type singularities. This information is captured by the index

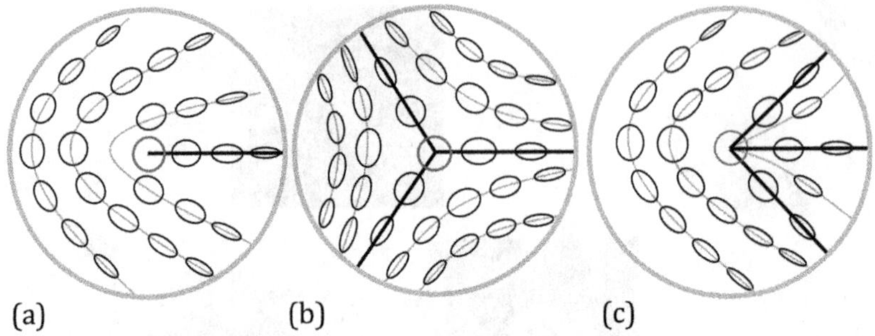

Figure 6.8. Morphology of C-type polarization singular structures. Here (a)–(c) show the lemon, star and monstar C-point singularities, respectively.

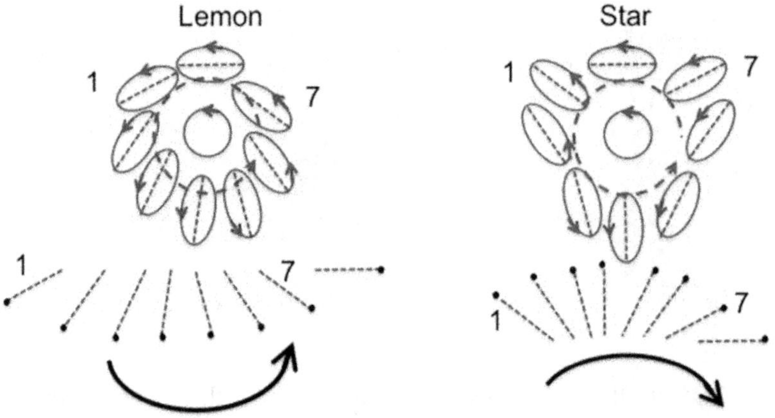

Figure 6.9. A Lemon C-point and a Star C-point. The sense of rotation of major axis of the neighboring SOPs around the circular polarization state is opposite in the two cases. The amount of rotation is $\pm\pi$ radians.

$$I_c = \frac{1}{2\pi} \oint \nabla\gamma \cdot d\mathbf{r}, \tag{6.10}$$

which is a line integral of gradient of γ (angle of orientation of the major axis of polarization ellipse) over a closed loop surrounding the C-point. As the γ is defined modulo π, the I_c is quantized in units of 1/2. Therefore, a C-point can be classified by its handedness (left-handed or right-handed) and index I_c. Lemon and monstar have $I_c = 1/2$ while star has $I_c = -1/2$. Monstar is a transitional singularity which only appears before creation or annihilation effects [23]. Monstar has the same index as a lemon and the same number of terminating lines as a star, which is why it is sometimes called a (le)monstar [27].

The vector-point singularities are isolated, stationary points of a linearly polarized vector field at which the orientation of the electric field vector is undefined. For a linearly polarized field, the handedness is undefined and for a V-point, even the orientation angle is undefined. Therefore, the field itself is zero at a V-point (note that this is not necessarily true for a C-point). The V-points are characterized using the Poincaré–Hopf index η which is given by the line-integral,

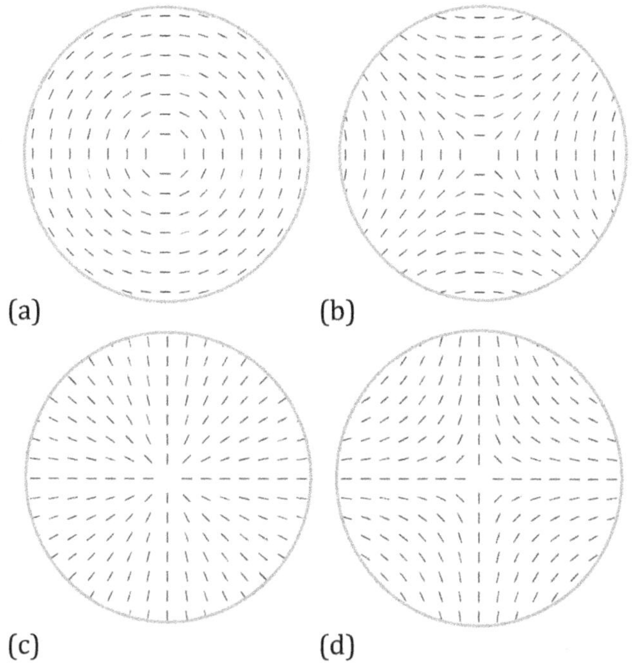

Figure 6.10. Morphology of V-type polarization singular structures. Here (a)–(d) show type I, type II, type III and type IV V-point singularities, respectively. Type I and type III are commonly known as azimuthal and radial polarizations.

$$\eta = \frac{1}{2\pi} \oint \nabla\gamma \cdot d\mathbf{r} \tag{6.11}$$

where the integral is evaluated over a closed loop surrounding the V-point. Azimuthally and radially polarized beams are lowest order V-point singular beams with $\eta = 1$, as shown in figures 6.10(a) and (c), respectively. The lowest order V-point singular beams with $\eta = -1$ are illustrated in figures 6.10(b) and (d).

6.3.3 Polarization singularities as vector superposition of OAM states

So far, we have explained polarization singularities in terms of the topological properties associated with the properties of polarization ellipses or orientations of linear polarization distributions. An explanation of this kind can sometimes leave a beginning reader somewhat confused as it is not clear how such states with exotic topological properties can be understood from a simpler picture of decomposing an arbitrary state as a superposition of two orthogonally polarized components. In circular basis representation of polarization states, the phase difference between orthogonal circular basis states decides the azimuth (γ) of the resulting polarization state. Hence to rotate the azimuth in the polarization distribution (like in figure 6.9), one has to have monotonically increasing or decreasing phase difference between circular basis components. Hence by having appropriate phase difference gradient in the component states, the required γ variation can be realized in the polarization distribution. For the azimuth γ to rotate in a clockwise direction the phase difference between the component beams should increase in a clockwise sense with respect to a point. One possibility to have such a phase difference is to superpose a negative charged vortex wave ($l = -1$) in RCP with a zero charge state ($l = 0$) in LCP. At a given radial distance from the vortex position, since the amplitude distributions in the two beams are not same, during superposition at each point (on a given circle) the amplitude difference between the two beams leads to elliptical polarization. At the same time the phase difference between the two beams will lead to different orientations of the polarization ellipses. At the location of the vortex beam, the amplitude is zero while at the same location in the zero charge wave the amplitude is not zero. This leads to circular polarization at the center in the superposition. The resulting polarization distribution is that of a C-point. Likewise, polarization singularities (C-points and V-points) can be seen as superposition of two beams with different OAM in orthogonal circular polarizations [28].

Let us consider the orthogonal basis elements corresponding to the left and right-circularly polarized ($\hat{e}_R$, $\hat{e}_L$) light where $\hat{e}_R = (\hat{x} - i\hat{y})/\sqrt{2}$ and $\hat{e}_L = (\hat{x} + i\hat{y})/\sqrt{2}$. The general equation for a vector beam (on ignoring the $\exp(ikz)$ dependence) is then given by expression,

$$\mathbf{E}(r, \theta, z) = E_R(r, \theta, z)\hat{e}_R + E_L(r, \theta, z)\hat{e}_L. \tag{6.12}$$

In order to introduce a polarization singularity in the vector beam, the two transverse components (E_R, E_L) are then taken to be the $LG(p, l)$ modes,

$$\mathbf{E}(x, y, z) = \frac{1}{\sqrt{2}}[LG(p_1, l_1)e^{i\delta_1}\hat{e}_R + LG(p_2, l_2)e^{i\delta_2}\hat{e}_L]. \tag{6.13}$$

As discussed earlier, the $LG(p, l)$ modes contain an on-axis phase singularity when $l \neq 0$. The C-point polarization singularities are generated when $l_1 \neq l_2$, whereas for V-point singularity the required condition is $p_1 = p_2$, $l_1 = -l_2$. The lowest order C-points ($l_1 = 1$, $l_2 = 0$, $\delta_1 = \delta_2 = 0$) and V-points ($l_1 = 1$, $l_2 = -1$, $\delta_1 = 0$) can be written for $p = 0$ as:

$$\mathbf{E}_C(x, y, z) = \frac{1}{\sqrt{2}}[LG(0, 1)\hat{e}_R + LG(0, 0)\hat{e}_L] \tag{6.14}$$

$$\mathbf{E}_V(x, y, z) = \frac{1}{\sqrt{2}}[LG(0, 1)\hat{e}_R + LG(0, -1)e^{i\delta_2}\hat{e}_L] \tag{6.15}$$

The polarization singularities have rich topological structures with their own index conservation laws that follow from the behavior of the OAM modes in the orthogonal polarization states. A detailed discussion of these properties may be found in [13]. We observe that a beam carrying a C-point singularity has a net non-zero OAM while the total OAM of a V-point carrying beam is zero. The decomposition of polarization singular beams in circular basis is illustrated in figure 6.11.

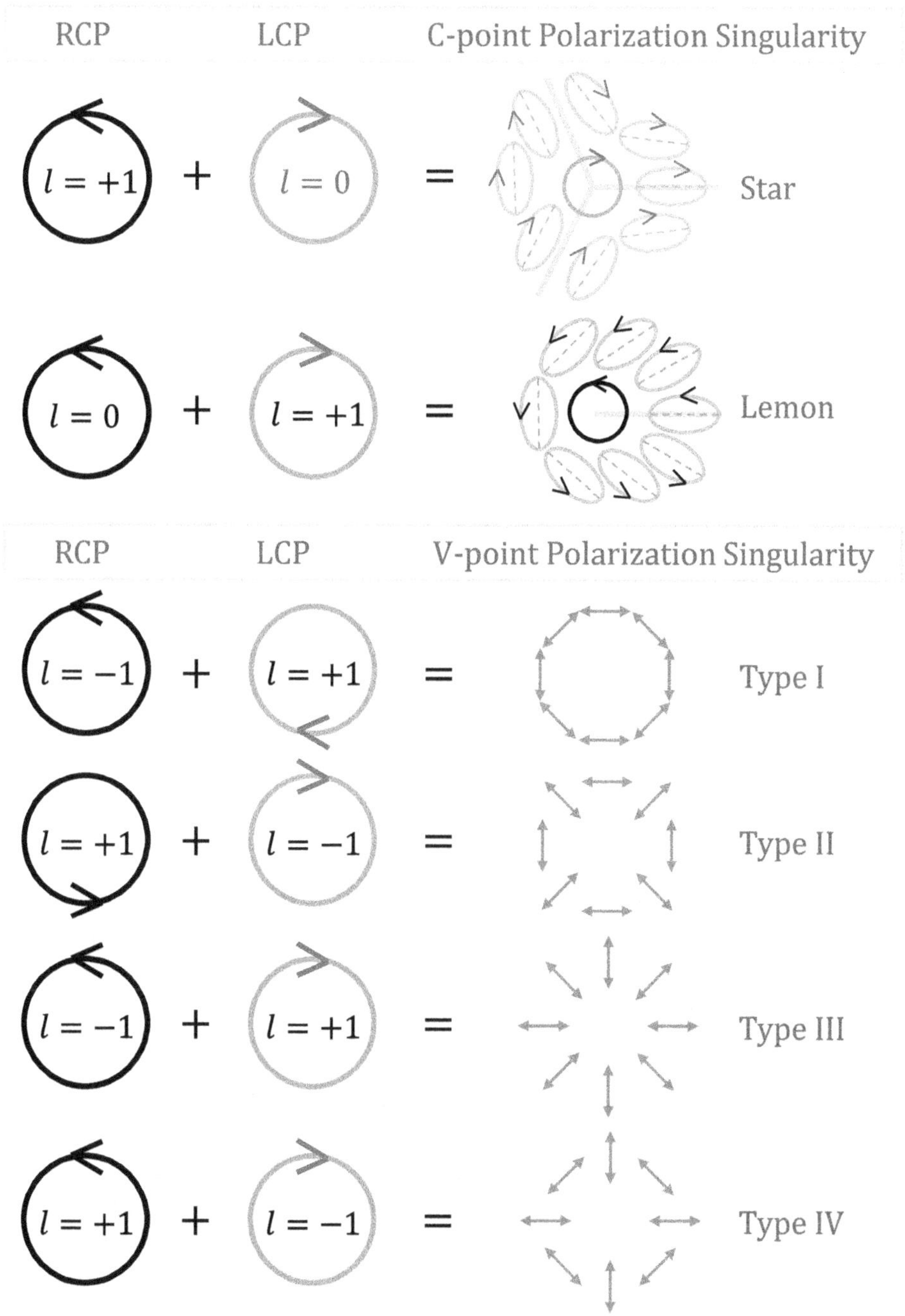

Figure 6.11. The figure illustrates the decomposition of C-point and V-point polarization singular structures in RCP-LCP ($\hat{e}_R - \hat{e}_L$) basis. Note that the right- and left-handedness is depicted by black and brown colors, respectively. The C-point structures shown here are a left-handed star and a right-handed lemon. The green color depicts linearly polarized light.

Table 6.2. Superposition in circular basis. RCP-V means vortex in right-circularly polarized light and LCP-V means vortex in left circular polarized light. $\Delta\phi$ is the constant phase shift given to one of the component beams. Note that in the superposition, generally when $l_1 \neq l_2$ and $l_1, l_2 \neq 0$, dark C-point results as the amplitude variation in the two polarizations are different, and the phase difference between them leads to rotating azimuth in the SOP distribution. Higher index C-points and V-points are also possible but are not tabulated here.

RCP-V (l_1)	LCP-V (l_2)	$\Delta\phi$	Polarization singularity	Type	Index
1	0	0	LH star	bright C-point	$I_C = -\frac{1}{2}$
0	1	0	RH lemon	bright C-point	$I_C = +\frac{1}{2}$
0	1	ϕ_0	RH lemon (rotated)	bright C-point	$I_C = +\frac{1}{2}$
−1	0	0	LH lemon	bright C-point	$I_C = +\frac{1}{2}$
−1	0	ϕ_0	LH lemon	bright C-point	$I_C = +\frac{1}{2}$
−1	+1	0	Radial polarization	V-point	$\eta = +1$
−1	+1	π	Azimuthal polarization	V-point	$\eta = +1$
1	−1	0	Type IV	V-point	$\eta = -1$
1	−1	π	Type II	V-point	$\eta = -1$
+1	+2	0	RH lemon	dark C-point	$I_C = +\frac{1}{2}$
0	2	0	RH (Radial) ellipses	C-point	$I_C = +1$

In circular basis we see that each superposition of OAM states leads to a polarization singularity, as given in table 6.2.

6.4 Stokes phase distribution and azimuth distribution

The polarization singularities can also be described using the concept of Stokes field. Using the Stokes parameters one may further define a mathematical construct called Stokes fields. For example, using Stokes parameters S_1 and S_2 a complex field, namely $S_{12} = S_1 + iS_2 = A_{12}\exp(\phi_{12})$, can be constructed [9, 10, 29, 30]. The Stokes phase $\phi_{12} = \arctan(S_2/S_1)$ can be seen to be equal to $\gamma/2$. Hence the phase vortices of the complex Stokes field S_{12} are the polarization singularities. Therefore, constructing a Stokes field from the measured Stokes parameters (each parameter is a spatial distribution for inhomogenously polarized beam), is helpful in identifying the polarization singularities. In figure 6.12(a) in the polarization distribution, the presence of two polarization singularities with opposite I_c index can be identified as two vortex phase distributions in the Stokes phase. A V-point singularity and its Stokes phase distribution are shown figure 6.12(b). However, the limitation in using Stokes fields is that the Stokes phase distribution does not distinguish the right- and left-handed C-point singularities and the dark or bright C-point singularities. Also, this picture does not distinguish between integer charged C-points and V-point singularities. These are illustrated in figure 6.12(c).

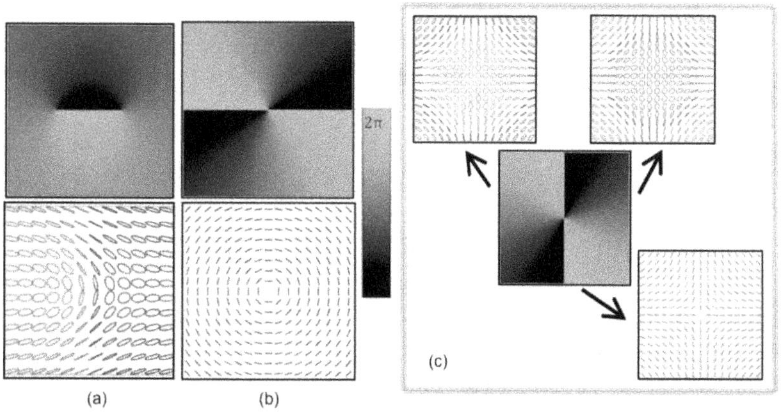

Figure 6.12. Stokes ϕ_{12} phase distributions and the corresponding SOP distributions are shown. Here, panels (a) and (b) represent a C-point dipole and V-point, respectively. Panel (c) shows that the Stokes phase is the same for two ellipse field singularities having $I_c = \pm 1$ right-handed, left-handed and for a vector field singularity with $\eta = -1$.

6.5 Generation and detection of polarization singularities

In the previous sections, the basics of polarization singularities have been explained. From the understanding of the decomposition of these singularities in orthogonal polarization bases, one can think of numerous techniques to realize them in the laboratory [4, 31–33]. By using a Mach–Zehnder, or a Michelson, or a Sagnac configuration, the incident light is split into two orthogonal polarization states and in either one or in both the beams, a phase vortex is inserted by using a vortex phase plate (or via a spatial light modulator). These two beams can then be combined back in circular basis to realize the polarization singularity. Another method is to use a single polarization element called S-wave plate or q-plates [34, 35]. These plates have spatially varying polarization characteristics. Consider a half-wave plate that can rotate of the azimuth of the incident beam's SOP. By aligning the fast/slow axis of the half-wave plate to different directions at different spatial locations, the incident homogenous light can be converted into a beam containing polarization singularity. Nano-structured materials [36] have also been designed to realize inhomogeneous polarization distributions that contain polarization singularities. Another interesting approach to generate polarization singularities uses stress engineered optics [37]. A schematic setup for the generation of polarization singularities is shown in figure 6.13. A 45° linearly polarized light is split equally at the polarizing beam splitter. Two spiral phase plates SPP1 and SPP2 inserted into the two arms of the interferometer implant a vortex in each of the beams, of charge say l_1 and l_2 in two beams, respectively. The second beam splitter combines the two beams such that the two vortex cores exactly coincide, and combined beams propagate collinearly. A quarter wave plate QWP oriented at 45° ensures that the superposition is in circular polarization basis. When $l_1 = -l_2$ V-point is generated; $l_1 = 0$ and $l_2 \neq 0$ or ($l_1 \neq 0$ and $l_2 = 0$) bright C-points are generated. When $l_1 \neq l_2$ and neither of them is of zero

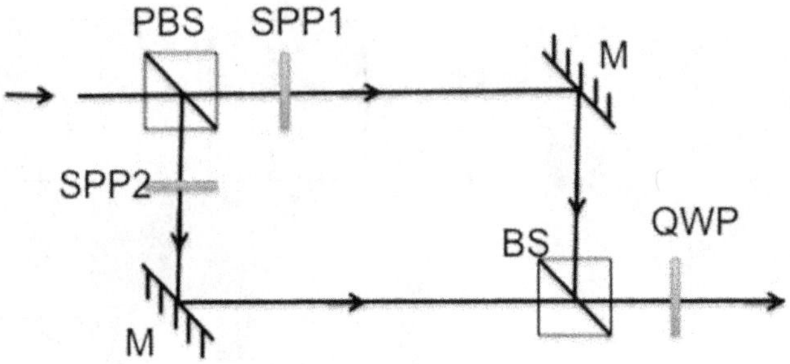

Figure 6.13. A schematic setup for the generation of C-point and V-point singularities.

charge, a dark C-point is generated. There are many methods based on interference [38, 39] and diffraction [40, 41] for the detection of phase singularities. But the number of methods available for the detection of polarization singularities are scarce at this point in time. One way to detect the polarization singularity is to obtain the spatially varying Stokes parameters using, say a Stokes camera, and analyze the distributions or by drawing SOP distributions. There is one method, based on interferometry to detect the polarization singularities involved [42]. Recently diffraction and polarization transformation-based methods [43] have been demonstrated for V-point detection.

6.6 Applications of polarization singular beams

Polarization singular beams have found some interesting applications. Radially polarized light has been shown experimentally and theoretically to focus to a spot which is significantly smaller than those achievable through linear or circular polarization states [44–47]. This effect is attributed to the strong longitudinal electric field component which is generated in the vicinity of the focus when a radially polarized light beam is passed through a high numerical aperture. In contrast, the azimuthally polarized light generates a purely transverse electric field which is zero at the center. Based on these unique properties of strong longitudinal electric field component and small focal spot size, radially polarized beams are being used in a variety of applications such as beam trapping, material processing (or laser-based cutting) and optical memories. Polarization-based spatial filtering can be done for directional and non-directional edge enhancement by making use of an s-wave plate [48]. The C-point polarization singular beams can be used to measure the optical activity in chiral samples [49] by detecting the singularity structure rotation that can be sensed by an analyzer. We will discuss later in this book how the beams containing C-point singularities are robust against intensity fluctuations on propagation through a random medium in comparison with scalar Gaussian beams.

References

[1] Hecht E and Zajac A 1974 *Optics* (Reading, MA: Addison-Wesley)
[2] Goldstein D H 2011 *Polarized Light* (Boca Raton, FL: CRC Press)
[3] Collett E 1993 *Polarized Light New York* (Marcel Dekker)

[4] Ruchi B, Bhargava Ram S and Senthilkumaran P 2017 Hopping induced inversions and Pancharatnam excursions of C-points *Opt. Lett.* **42** 4159–62

[5] Maurer C, Jesacher A, Furhapter S, Bernet S and Marte M R 2007 Tailoring of arbitrary optical vector beams *New J. Phys.* **9** 78

[6] Pal S K, Ruchi and Senthilkumaran P 2017 Polarization singularity index sign inversion by half-wave plate *Appl. Opt.* **56** 6181–9

[7] Cameron R P, Barnett S M and Yao A M 2012 Optical helicity, optical spin and related quantities in electromagnetic theory *New J. Phys.* **14** 053050

[8] Cameron R P, Barnett S M and Yao A M 2014 Optical helicity of interfering waves *J. Mod. Opt.* **61** 25–31

[9] Dennis M R 2002 Polarization singularities in paraxial vector fields: morphology and statistics *Opt. Commun.* **213** 201–21

[10] Freund I 2002 Polarization singularity indices in Gaussian laser beams *Opt. Commun.* **201** 251–70

[11] Freund I, Soskin M S and Mokhun A I 2002 Elliptic critical points in paraxial optical fields *Opt. Commun.* **208** 223–53

[12] Rosales-Guzman C, Ndagano B and Forbes A 2018 A review of complex vector light fields and their applications *J. Opt.* **20** 123001 11

[13] Senthilkumaran P 2018 *Singularities in Physics and Engineering Properties: Methods and applications*IOP Series in Advances in Optics, Photonics and Optoelectronics *(Bristol: IOP Publisihg)*

[14] Ghai D P, Senthilkumaran P and Sirohi R S 2008 Shearograms of optical phase singularity *Opt. Commun.* **281** 1315–22

[15] Roux F S 2003 Optical vortex density limitation *Opt. Commun.* **223** 31–7

[16] Nye J F and Berry M V 1974 Dislocations in wave trains *Proc. R. Soc.* **336** 165–90

[17] Basisty I V, Soskin M S and Vasnetsov M V 1993 Optics of light beams with screw dislocations *Opt. Commun.* **103** 422–8

[18] Basisty I V, Soskin M S and Vasnetsov M V 1995 Optical wavefront dislocations and their properties *Opt. Commun.* **119** 604–12

[19] Allen L, Beijersbergen M W, Spreeuw R J and Woerdman J P 1992 Orbital angular momentum of light and the transformation of Laguerre–Gaussian laser modes *Phys. Rev.* A **45** 8185–9

[20] Allen L, Padgett M J and Babikar M 1999 *The orbital angular momentum of light Progress in Optics* **vol 39** (Amsterdam: Elsevier)

[21] van Enk S J and Nienhuis G 1992 Eigenfunction description of laser beams and orbital angular momentum of light *Opt. Commun.* **94** 147–58

[22] Nienhuis G and Allen L 1993 Paraxial wave optics and harmonic oscillators *Phys. Rev.* A **48** 656–65

[23] Gbur G 2015 *Singular Optics* (Boca Raton, FL: CRC Press)

[24] Freund I 2002 Polarization singularity indices in Gaussian laser beams *Opt. Commun.* **201** 251–70

[25] Dennis M R 2002 Polarization singularities in paraxial vector fields: morphology and statistics *Opt. Commun.* **213** 201–21

[26] Freund I 2001 Polarization flowers *Opt. Commun.* **199** 47–63

[27] Dennis M R, O'Holleran K and Padgett M J 2009 Optical vortices and polarization singularities *Prog. Opt. (Ed: E. Wolf)* **53** 293–363

[28] Senthilkumaran R P and Pal S K 2020 Phase singularities to polarization singularities *Int. J. Opt.* **2020** 2812803
[29] Pal R S K and Senthilkumaran P 2017 C-point and V-point singularity lattice formation and index sign conversion methods *Opt. Commun.* **393** 156–68
[30] Pal S K and Senthilkumaran P 2019 Synthesis of stokes singularities *Opt. Lett.* **44** 130–3
[31] Tidwell W C, Ford D H and Kimura W D 1990 Generating radially polarized beams interferometrically *Appl. Opt.* **29** 2234–9
[32] Aadhi A, Vaity P, Chithrabhanu P, Redday S G, Prabhakar S and Singh R P 2016 Non-coaxial superposition of vector vortex beams *Appl. Opt.* **55** 1107–11
[33] Bhargava Ram B S, Sharma A and Senthilkumaran P 2017 Diffraction of V-point singularities through triangular apertures *Opt. Express* **25** 10270–5
[34] Machavariani G, Lumer Y, Moshe I, Meir A and Jackel S 2008 Spatially variable retardation plate for efficent generation of radially and azimuthally polarized beams *Opt. Commun.* **281** 732–8
[35] Marucci L, Manzo C and Paparo D 2006 Optical spin-to-orbital angular momentum conversion in inhomogeneous anisotropic media *Phys. Rev. Lett.* **96** 1639905
[36] Biener G, Niv A, Kleiner V and Hasman E 2002 Formation of helical beams by use of Pancharatnam-Berry phase optical elements *Opt. Lett.* **27** 1875–7
[37] Liang K, Ariyawansa A and Brown T G 2019 Polarization singularities in a stress-engineered optic *J. Opt. Soc. Am.* A **36** 312–9
[38] Vaughan J M V and Willetts D V 1983 Temporal and interference fringe analysis of excimer tem_{01}* laser *J. Opt. Soc. Am.* **73** 1018–21
[39] Ghai D P, Senthilkumaran P and Sirohi R S 2008 Detection of phase singularity using a lateral shear interferometer *Opt. Laser Engg.* **46** 419–23
[40] Moreno I, Davis J A, Melvin B, Pascoguin L, Mitry M J and Cottrell D M 2009 Vortex sensing diffraction gratings *Opt. Lett.* **34** 2927–9
[41] Ghai D P, Senthilkumaran P and Sirohi R S 2008 Single slit diffraction of an optical beam with phase singularity *Opt. Laser Engg.* **47** 123–6
[42] Angelsky O V, Mokhun I I, Mokhun A I and Soskin M S 2002 Interferometric methods in diagnostics of polarization singularities *Phys. Rev.* E **65** 1–5
[43] BhargavaRam B S, Sharma A and Senthilkumaran P 2017 Probing the degenerate states of V-point singularities *Opt. Lett.* **42** 3570–3
[44] Dorn R, Quabis S and Leuchs G 2003 Sharper focus for a radially polarized light beam *Phys. Rev. Lett.* **91** 233901 12
[45] Quabis S, Dorn R, Eberler M, Glockl O and Leuchs G 2000 Focusing light to a tighter spot *Opt. Commun.* **179** 1–7
[46] Lerman G M and Levy U 2008 Effect of radial polarization and apodization on spot size under tight focusing conditions *Opt. Express* **16** 4567–81
[47] Zhan Q and Leger J R 2002 Focus shaping using cylindrical vector beams *Opt. Express* **10** 324–31
[48] Bhargava Ram B S, Senthilkumaran P and Sharma A 2017 Polarization-based spatial filtering for directional and nondirectional edge enhancement using an s-waveplate *Appl. Opt.* **56** 3171–8
[49] Samlan C T, Suna R R, Naik D N and Viswanathan N K 2018 Spin-orbit beams for optical chirality measurement *Appl. Phys. Lett.* **112** 031101

Chapter 7

Theory of wave propagation in a turbulent medium

With the discussion of basic mathematical tools and description of OAM states of light as well as polarization singularities, we are now ready to tackle the problem of sending light beams through randomly fluctuating media like atmospheric turbulence. Long-range beam propagation through turbulence is of interest to defense as well as free-space optical communication (both classical and quantum) systems. While the study of laser beam propagation through the atmosphere has a long history of several decades, we believe that these studies have mainly focused on scalar beams. Secondly, there is a general lack of easily accessible but fairly rigorous beam propagation computation tools. The remaining part of this book will therefore first aim to explain the basic theory of light propagation through turbulence.

7.1 Electromagnetic wave equation in a random medium

The nature of propagation of monochromatic electromagnetic fields in free space was already described in chapter 3. A turbulent medium like the atmosphere is characterized by spatially randomly varying dielectric constant $\varepsilon(r, t)$. We assume that the atmosphere has constant magnetic permeability and zero conductivity implying an absence of free charges and external currents. We start with equation (3.19) which describes the propagation of the E-field in a medium with spatially varying refractive index.

$$\nabla^2 \boldsymbol{E} + \nabla(\boldsymbol{E}.\ \nabla \ln \varepsilon) = \mu \frac{\partial^2}{\partial t^2}(\varepsilon \boldsymbol{E}) \tag{7.1}$$

The second term on the left-hand side represents the coupling between the orthogonal polarizations of the electric field on propagation through a random medium. We will calculate the magnitude of the depolarization term in the next section. For line-of-sight propagation, the magnitude of this term is very small at

doi:10.1088/978-0-7503-5959-7ch7

optical wavelengths and this term can be safely ignored when long-range turbulence propagation is of interest. Let us assume that the time dependence of the electromagnetic fields is sinusoidal of the form $\exp(-i\omega t)$ where ω is the frequency of the wave, so that the fields can be written as:

$$\boldsymbol{E}(\boldsymbol{r}, t) = \boldsymbol{E}_o(\boldsymbol{r})\exp(-i\omega t) \tag{7.2}$$

$$\boldsymbol{B}(\boldsymbol{r}, t) = \boldsymbol{B}_o(\boldsymbol{r})\exp(-i\omega t) \tag{7.3}$$

Usually the dielectric constant $\varepsilon(r, t)$ is expressed as:

$$\varepsilon(\boldsymbol{r}, t) = \langle\varepsilon(\boldsymbol{r})\rangle + \varepsilon_1(\boldsymbol{r}, t) \tag{7.4}$$

where the $\langle\varepsilon\rangle$ is the deterministic (non-random) average value of dielectric constant which can be a function of position. The random fluctuations about the mean value are contained in the term ε_1 which varies with both space and time. By definition, $\langle\varepsilon_1\rangle = 0$ and $\langle|\varepsilon_1|\rangle \ll \langle\varepsilon\rangle$. This condition is valid, for example in the troposphere where $\langle\varepsilon\rangle \approx 1$ and $\langle|\varepsilon_1|\rangle \sim 10^{-6} - 10^{-4}$. The dielectric constant can also be written in terms of the random refractive index function $n(\boldsymbol{r}, t)$ as:

$$\varepsilon(\boldsymbol{r}, t) = \varepsilon_o n^2(\boldsymbol{r}, t) \tag{7.5}$$

where $n(\boldsymbol{r}, t)$ is also a random function of space and time, defined in a similar fashion as equation (7.4):

$$n(r, t) = \sqrt{\frac{\varepsilon(\boldsymbol{r}, t)}{\varepsilon_o}} = \langle n(\boldsymbol{r})\rangle + n_1(r, t) \tag{7.6}$$

where $\langle n(\boldsymbol{r})\rangle$ gives the average value of the refractive index and $n_1(\boldsymbol{r}, t)$ gives the fluctuations about the mean value. For small fluctuations, we get $\varepsilon_1(\boldsymbol{r}, t) \approx 2n_1(\boldsymbol{r}, t)$. For the moment, we will suppress the time dependence of the dielectric constant and only pay attention to its spatial properties, therefore assuming ε_1 to be independent of time, i.e., $\varepsilon(\boldsymbol{r}) = \langle\varepsilon(\boldsymbol{r})\rangle + \varepsilon_1(\boldsymbol{r})$. Such an assumption is valid as the atmospheric fluctuations are much slower compared to the electromagnetic frequencies. These fluctuations are induced by two distinct processes: (1) internal rearrangement of the turbulent eddies, and (2) due to the air stream which carries these eddies. If the velocity of the air stream which carries these eddies is very much greater than the turbulent velocity with which the eddies are mixing, then the structure of the turbulent air does not change and is simply carried forward along with the wind. This is known as the *Taylor's frozen flow hypothesis* and is usually invoked whenever temporal fluctuations of the atmosphere are to be studied. This allows us to take ε out of the time derivative in equation (7.1) and we get:

$$\nabla^2\boldsymbol{E}_o(\boldsymbol{r}) + \frac{\omega^2 n^2(\boldsymbol{r})}{c^2}\boldsymbol{E}_o(\boldsymbol{r}) = 0 \tag{7.7}$$

This stochastic wave equation differs from the conventional wave equation through the fact that now the refractive index is a random function of $\boldsymbol{r}$. Therefore, the equation is a homogeneous partial differential equation with space dependent coefficients. The refractive index function $n(\mathbf{r})$ varies from point to point in an unpredictable manner making it impractical to describe its value at all points in space and so it becomes imperative to describe the medium statistically. However, even for a given statistical description of $n(\mathbf{r})$, the exact solution for the wave equation is difficult to obtain and one has to apply approximate methods to get a practical solution. Just like in any other problem containing random variables, one can only attempt to solve equation (7.7) to obtain statistical average (or mean value) for the desired quantity of interest, for example, mean irradiance, average beam width, the variance of angle of arrival etc. The three components of the electric field vector in equation (7.7) can be solved independently from each other as we have already dropped the depolarization term which was responsible for the coupling between the three components. Therefore, we can decompose the equation (7.7) into three scalar equations for each component and ignore the vector nature of the electric field for the time being. Let $U(\mathbf{r})$ represent one of the scalar components transverse to the direction of propagation, then we can write equation (7.7) in the form commonly known as the *stochastic Helmholtz equation*:

$$\nabla^2 U(\mathbf{r}) + k^2 n^2(\mathbf{r}) U(\mathbf{r}) = 0 \tag{7.8}$$

where $k = \omega/c = 2\pi/\lambda$ is the wave number. Before we proceed to look into the various theoretical methods that have been employed to solve the above equation, we must first be able to provide the statistical description of the refractive index fluctuations denoted by $n_1(\mathbf{r})$. Early attempts to solve the stochastic Helmholtz equation were based on the geometrical optics method. In the geometrical optics method, diffraction effects are ignored and the phase fluctuations are calculated as per Snell's law for when a ray passes through regions of different refractive index variations. The curvature of the turbulent eddies causes focusing or de-focusing of rays as if they are passing through converging or diverging lenses which results in amplitude fluctuations. The results of this method are valid when the width of the scattering cone is smaller than the size of a turbulent eddy. This is usually stated in terms of the Fresnel length $\sqrt{\lambda L}$ (where L denotes the propagation distance) which should be smaller than the size of the turbulent eddies. Otherwise the diffraction effects can become important and the amplitude fluctuations are not correctly predicted [1]. Before providing details of models for light propagation through turbulence, it is important to understand the nature and physical origins of refractive index fluctuations in the atmosphere that we will discuss in the next section.

7.2 Description of the refractive index fluctuations in the atmosphere

While it is common knowledge that refractive index of the atmosphere fluctuates in time, the nature of these fluctuations, their space–timescales and the magnitude of index fluctuations is important to understand before we can begin to describe sophisticated models of beam propagation in turbulence.

7.2.1 Origin of fluctuations in the index of refraction

The solar energy during daytime heats up the Earth's surface along with the surrounding air. It is well known that the air closer to the ground becomes warmer due to additional radiative heat transfer from the Earth's surface forming a vertical temperature gradient in the atmosphere. The hotter air is lighter or less dense than the cool air. This results in a rapid ascent of the hotter air into the atmosphere while the surrounding cooler air descends to take its place. This turbulent mixing of air in the atmosphere breaks down large air masses into randomly distributed pockets, each with its own characteristic temperature. The local index of refraction of air being sensitive to temperature, develops a random profile.

The turbulent mixing of the air gives rise to randomly distributed regions of high or low refractive index which are known as turbulent eddies. The eddies can be thought of as random lenses having different shapes and spatial sizes which are randomly moving in the atmosphere. However, the refractive index is also varying within an eddy as there are usually no refractive index discontinuities present in the atmosphere. The refractive index changes due to atmospheric turbulence are sufficiently small and the light beam passing through the turbulent eddies can be safely assumed to experience only phase changes with no accompanying amplitude changes. The phase fluctuations experienced by the input light beam are random and can give rise to focusing or de-focusing effects and beam wandering effects over large propagation distances. The phase fluctuations accumulated by the beam wavefront result in random destructive and constructive interference and on propagation through multiple eddies give rise to fluctuations in the intensity profile of the beam, an effect commonly known as beam scintillation.

Kolmogorov's theory of atmospheric turbulence draws from similarities from fluid dynamics [2]. In the study of motion of a viscous fluid, two states can be observed—laminar and turbulent. Reynolds used a non-dimensional quantity known as Reynolds number, $\mathrm{Re} = v_{\mathrm{avg}} l_s / k_v$ to characterize these two types of fluid motion. Here, v_{avg} is the average velocity of the viscous fluid and l_s denotes the characteristic dimension (or size) of the flow. The kinematic viscosity of the fluid is given by k_v. The fluid flow changes states from laminar to turbulent as the average velocity of the flow is increased. In laminar flow, the fluid follows a smooth path and the velocity characteristics are uniform or change in some deterministic fashion. However, when Reynolds number exceeds a critical value, the flow becomes turbulent and dynamic mixing of the fluid occurs leading to formation of turbulent eddies. For atmosphere, taking $v_{\mathrm{avg}} = 2\ \mathrm{m\ s^{-1}}$, $l_s = 10\ \mathrm{m}$ and $k_v = 1.5 \times 10^{-5}\ \mathrm{m^2\ s^{-1}}$, the Reynolds number is approximately given by $\mathrm{Re} = 1.3 \times 10^6$ which is very high (for laminar flow, the Reynolds number is typically up to order of 10^3), and so the atmosphere shows turbulent air mixing.

In order to model optical turbulence, it is important to describe the statistics of the turbulent eddies. The result of the turbulent mixing of air is that the energy injected into large-scale air masses is transferred to air masses on smaller scales. Richardson developed the *energy cascade theory* to visualize this energy transfer, as shown in figure 7.1 [3, 4]. As the larger eddies break into small sized eddies, there is a

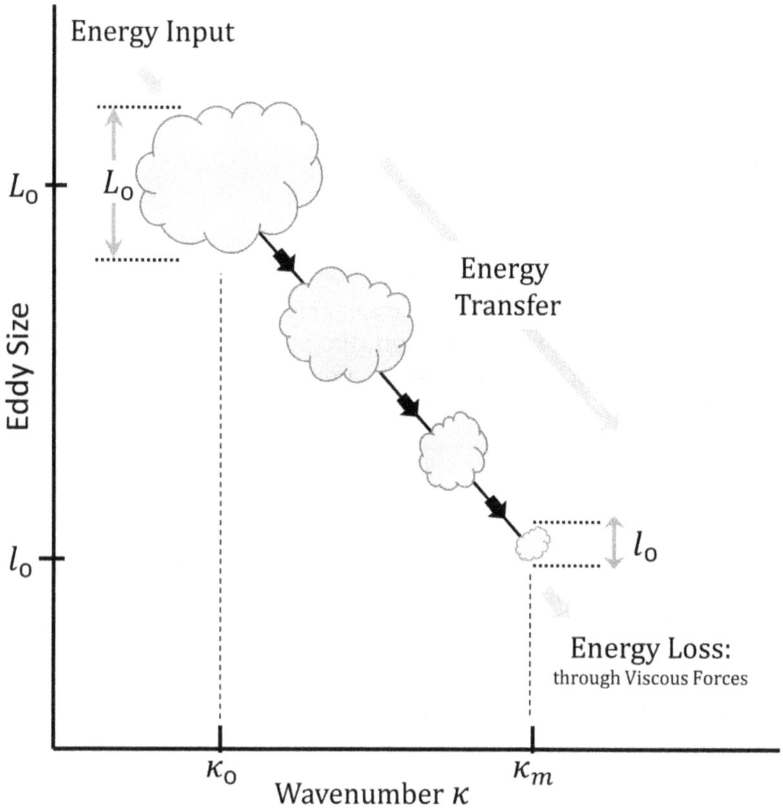

Figure 7.1. Energy cascade theory of turbulence: the large sized eddies subdivide into smaller and smaller eddies until they completely disappear. The homogeneous and isotropic turbulence is characterized by eddies with sizes lying between L_o and l_o which denote the outer and inner scales of turbulence, respectively.

continuum of eddy sizes. The eddies which are treated as homogeneous and isotropic are bounded by the eddy sizes L_o and l_o. These sizes denote the outer and inner scale of the atmospheric turbulence. Eddies of sizes larger than L_o are no longer isotropic and their structure is not well defined while eddies of sizes smaller than l_o lose most of their energy in heat through viscous dissipation processes. As the eddies become smaller and smaller, the relative amount of energy injected and energy dissipated becomes equal and so they lose much of their kinetic energy, thus decreasing the Reynolds number to the order of unity. The inner scale l_o is typically of the order of millimeters near the ground while the outer scale L_o can take values in meters. Both quantities increase as one moves vertically away from the ground.

At optical frequencies, the refractive index of the atmosphere is known to depend on the local pressure and temperature through the relation [3, 5]:

$$n_1 = 77.6[1 + 7.52 \times 10^{-3}\lambda^{-2}]\frac{P}{T} \times 10^{-6} \tag{7.9}$$

where T and P denote the temperature and pressure of the air in Kelvins and millibars, respectively. Pressure fluctuations measured at a point with respect to the ground are relatively small and as a result, the temperature fluctuations play a major role in producing refractive index variations.

7.2.2 Spatial statistics of refractive index fluctuations

Correlation function and power-spectral density

The fluctuating part of the refractive index given by $n_1(\boldsymbol{r})$ is a random process as given by equation (7.6). The random process $n_1(\boldsymbol{r})$ is usually assumed to be stationary and statistically homogeneous. As a result, the auto-correlation of $n_1(\boldsymbol{r})$, given by $\Gamma_n(\boldsymbol{r}_1, \boldsymbol{r}_2)$ depends only on the spatial separation $\boldsymbol{r} = \boldsymbol{r}_1 - \boldsymbol{r}_2$ and not on the exact spatial positions $\boldsymbol{r}_1$ and $\boldsymbol{r}_2$ [6]. Therefore we write[1],

$$\begin{aligned} \Gamma_n(\boldsymbol{r}_1, \boldsymbol{r}_2) &= \langle n_1(\boldsymbol{r}_1) n_1(\boldsymbol{r}_2)\rangle \\ \Gamma_n(\boldsymbol{r}) &= \langle n_1(\boldsymbol{r}_1) n_1(\boldsymbol{r}_1 - \boldsymbol{r})\rangle \end{aligned} \tag{7.10}$$

The statistical distribution of the size and number of the turbulent eddies is characterized by the spatial power-spectral density (PSD) of $n_1(\boldsymbol{r})$, denoted as $\Phi_n(\boldsymbol{\kappa})$. Here $\boldsymbol{\kappa}$ is the wave number vector with orthogonal components: $\kappa_x = 2\pi f_x$, $\kappa_y = 2\pi f_y$ and $\kappa_z = 2\pi f_z$ and $|\boldsymbol{\kappa}| = \sqrt{\kappa_x^2 + \kappa_y^2 + \kappa_z^2}$. The auto-correlation function $\Gamma_n(\boldsymbol{r})$ and the power-spectral density $\Phi_n(\boldsymbol{\kappa})$ are related by the Weiner–Khinchine theorem (see chapter 2) and form a three-dimensional Fourier transform pair [6]:

$$\Gamma_n(\boldsymbol{r}) = \iiint_{-\infty}^{+\infty} \Phi_n(\boldsymbol{\kappa}) \exp(i\boldsymbol{\kappa} \cdot \boldsymbol{r})\, d\boldsymbol{\kappa} \tag{7.11}$$

$$\Phi_n(\boldsymbol{\kappa}) = \left(\frac{1}{2\pi}\right)^3 \iiint_{-\infty}^{+\infty} \Gamma_n(\boldsymbol{r}) \exp(-i\boldsymbol{\kappa} \cdot \boldsymbol{r})\, d\boldsymbol{r} \tag{7.12}$$

The power-spectral density $\Phi_n(\boldsymbol{\kappa})$ is considered as a measure of the relative abundance of the turbulent eddies with scale scales $l_x = 2\pi/\kappa_x$, $l_y = 2\pi/\kappa_y$ and $l_z = 2\pi/\kappa_z$. For the case of homogeneous and isotropic turbulence, the auto-correlation of $n_1(\boldsymbol{r})$ is spherically symmetric. Hence, $\Gamma_n(\boldsymbol{r})$ is only a function of the scalar distance $r = |\boldsymbol{r}|$, while $\Phi_n(\boldsymbol{\kappa})$ is a function of the scalar wave number κ which is related to the isotropic scale size $l = 2\pi/\kappa$. Both auto-correlation and PSD for homogeneous random processes are even functions [3, 7], allowing us to rewrite equations (7.11) and (7.12) as,

$$\Gamma_n(r) = \frac{4\pi}{r} \int_0^{+\infty} \kappa d\kappa \;\; \Phi_n(\kappa) \sin(\kappa r), \tag{7.13}$$

[1] In the case of any complex stationary random process $x(t)$, the correlation function is defined by $\Gamma_X(t_1, t_2) = \langle x(t_1) x^*(t_2)\rangle$, where the asterisk * denotes the complex conjugate.

$$\Phi_n(\kappa) = \frac{1}{2\pi^2\kappa}\int_0^{+\infty} rdr\ \Gamma_n(r)\sin(\kappa r). \tag{7.14}$$

Covariance function

The spatial covariance function of the refractive index field is defined by the ensemble average,

$$B_n(\boldsymbol{r}_1, \boldsymbol{r}_2) = \langle[n_1(\boldsymbol{r}_1) - \langle n_1(\boldsymbol{r}_1)\rangle][n_1(\boldsymbol{r}_2) - \langle n_1(\boldsymbol{r}_2)\rangle]\rangle \tag{7.15}$$

For statistically homogeneous turbulence, the moments of the refractive index fluctuations will be invariant under spatial translation. Then the mean value $\langle n_1(\boldsymbol{r})\rangle = m$ has some constant value, independent of the spatial coordinate r. The covariance function is now only dependent on the spatial separation $\boldsymbol{r} = \boldsymbol{r}_1 - \boldsymbol{r}_2$,

$$B_n(\boldsymbol{r}) = \langle[n_1(\boldsymbol{r}_1)\, n_1(\boldsymbol{r}_1 + \boldsymbol{r})]\rangle - |m|^2 \tag{7.16}$$

For refractive index fluctuations, $\langle n_1(\boldsymbol{r})\rangle = 0$, therefore, the covariance function is the same as the auto-correlation function.

Structure function

The mean value of refractive index at a given location $\boldsymbol{r}$ is not a constant and fluctuates over reasonably long length scales. As a result the random process $n_1(\boldsymbol{r})$ cannot be considered as a homogeneous (or spatially stationary) process. In this case, rather than looking at the random process $n_1(\boldsymbol{r})$ itself, one can define another random process $n_1(\boldsymbol{r} + \boldsymbol{r}') - n_1(\boldsymbol{r})$ which behaves very similarly to a stationary random process with slowly varying mean. For such processes, the structure function is a valuable statistical descriptor and is defined as [3, 8],

$$D_n(\boldsymbol{r}_1, \boldsymbol{r}_2) = \langle[n_1(\boldsymbol{r}_1) - n_1(\boldsymbol{r}_2)]^2\rangle \tag{7.17}$$

The subtraction process in the structure function removes the effect of slowly varying large-scale fluctuations as the difference operation acts like a high-pass filter. For a stationary random process, the structure function is related to the auto-correlation function $\Gamma_n(\boldsymbol{r})$ as,

$$D_n(\boldsymbol{r}) = 2[\Gamma_n(0) - \Gamma_n(\boldsymbol{r})] \tag{7.18}$$

Using equation (7.11), the structure function can be written in terms of the PSD as [3]:

$$D_n(\boldsymbol{r}) = 2\iiint_{-\infty}^{+\infty} [1 - \exp(i\boldsymbol{\kappa}\cdot\boldsymbol{r})]\ \Phi_n(\boldsymbol{\kappa})\ d\boldsymbol{\kappa} \tag{7.19}$$

For isotropic turbulence, $D_n(\boldsymbol{r})$ is a function of scalar distance r, so we may further obtain the following simplified expression:

$$D_n(r) = 8\pi\int_0^{\infty} d\kappa\ \Phi_n(\kappa)\left[1 - \frac{\sin(\kappa r)}{\kappa r}\right]\kappa^2 \tag{7.20}$$

We will have occasion to use the structure function in the next chapter when understanding the nature of phase screens generated by the fast Fourier transform method for simulating atmospheric turbulence.

7.2.3 Temporal evolution of the fluctuations

The atmosphere is a dynamically evolving random medium where the distribution of the turbulent eddies changes over millisecond timescales. The temporal fluctuations in the refractive index of the atmosphere play an important role in applications like imaging. As discussed previously, the effect of the time domain fluctuations are modeled using the Taylor's frozen flow hypothesis. This hypothesis treats the entire arrangement of turbulent eddies to be frozen during the measurement interval. The complete frozen air mass only undergoes a horizontal translation in the transverse plane due to the wind, as shown in figure 7.2. The transverse velocity of the wind $\boldsymbol{v}$ is assumed to be constant at a given location. Here, the transverse velocity refers to the component of the wind velocity normal to the observation path. The refractive index fluctuation at some time $t_2 > t_1$ is thus related to the fluctuation at t_1 by [7],

$$n_1(\boldsymbol{r}, t_2) = n_1(\boldsymbol{r} - \boldsymbol{v}(t_2 - t_1), t_1) \tag{7.21}$$

Therefore, the space–time covariance function becomes,

$$\Gamma_n^S(\boldsymbol{r}_1, t_1; \boldsymbol{r}_2, t_2) = \langle n_1(\boldsymbol{r}_1, t_1) n_1(\boldsymbol{r}_2, t_2) \rangle \tag{7.22}$$

$$= \Gamma_n(\boldsymbol{r}_1 - \boldsymbol{r}_2 + \boldsymbol{v}(t_2 - t_1)) \tag{7.23}$$

In the above expression, the fluctuations in the wind velocity have been neglected by assuming that the magnitude of velocity fluctuations is much less than the magnitude of the wind velocity. However, this might not be always true for cases when the wind flows near parallel to the line-of-sight of the optical system. The Taylor hypothesis also breaks down for large time intervals as the turbulent eddies ultimately get rearranged over time.

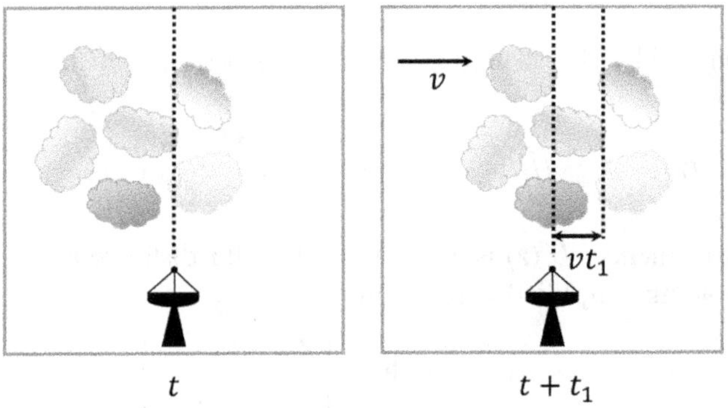

Figure 7.2. Taylor's hypothesis: frozen turbulent eddies translated by the wind's velocity over a time duration t_1.

7.2.4 Different models for power-spectral density of the refractive index fluctuations

In this section, we will briefly discuss the important PSD models. As discussed in section 7.2.1, the PSD is only described for the wave numbers which fall inside the inertial sub-range which is bounded by l_o and L_o where the PSD can be assumed to be isotropic. For wave numbers lying outside this range, the spectrum is generally considered to be anisotropic and does not have any closed form expression.

1. **Kolmogorov spectrum**

   The Kolmogorov theory predicts the PSD $\Phi_n(\kappa)$ to be of the form [4]:

   $$\Phi_n^K(\kappa) = 0.033\ C_n^2\ \kappa^{-11/3}\ \text{for}\ \frac{2\pi}{L_o} < \kappa < \frac{2\pi}{l_o} \tag{7.24}$$

   The quantity C_n^2 is called the structure constant of the refractive index fluctuations with units $m^{-2/3}$ which characterizes the strength of the refractive index fluctuations. In the limit $\kappa \to 0$, the expression for the spectrum $\Phi_n(\kappa)$ has a singularity in the form of a non-integrable pole. To overcome this difficulty and to incorporate the effect of inner and outer scales, alternate forms of PSD have been proposed.

2. **Tatarskii spectrum**

   As discussed earlier, the turbulent Rayleigh range eddies of sizes $l < l_o$ form the viscous dissipation range where the energy is mostly dissipated through viscosity effects in the form of heat. Tatarskii introduced a new model which truncates the spectrum in this dissipation range much more sharply than the $\kappa^{-11/3}$ form given by Kolmogorov [4]. The Tatarskii model for the PSD is given by:

   $$\Phi_n^T(\kappa) = 0.033\ C_n^2\ \kappa^{-11/3} \exp\left(-\frac{\kappa^2}{\kappa_m^2}\right)\ \text{for}\ \kappa \gg \frac{2\pi}{L_o} \tag{7.25}$$

   where $\kappa_m = 5.92/l_o$. The Tatarskii spectrum also contains a singularity when $L_o \to \infty$. Both the Kolmogorov and Tatarskii spectra are further modified so that they remain isotropic and finite for $\kappa < 1/L_o$. This brings us to the von Kármán spectrum.

3. **von Kármán spectrum**

   The Kolmogorov spectrum can be modified [4] to include wave numbers $\kappa < 1/L_o$ as,

   $$\Phi_n^V(\kappa) = \frac{0.033\ C_n^2}{(\kappa^2 + \kappa_o^2)^{11/6}}\ \text{for}\ 0 \leqslant \kappa \ll \frac{2\pi}{l_o} \tag{7.26}$$

   where $\kappa_o = 2\pi/L_o$. Similarly, the Tatarskii spectrum can be modified,

   $$\Phi_n^V(\kappa) = \frac{0.033 C_n^2}{(\kappa^2 + \kappa_o^2)^{11/6}} \exp\left(-\frac{\kappa^2}{\kappa_m^2}\right)\ \text{for}\ 0 \leqslant \kappa < \infty. \tag{7.27}$$

Equations (7.26) and (7.27) are known as the von Kármán spectrum and the modified von Kármán spectrum, respectively. Notice that a non-zero κ_o results in a finite value of PSD at $\kappa = 0$ while a non-zero κ_m rapidly brings the PSD to zero for $\kappa > \kappa_m$. The PSD in equation (7.27) is defined for the entire range of κ values, however, its form in the range $\kappa < 2\pi/L_o$ is an approximation.

The three spectra given by equations (7.24), (7.26) and (7.27) are shown in figure 7.3. The three distinct stages of turbulence process—the energy injection, inertial range and the energy dissipation are also indicated in the plot. Experimental measurements of the power spectrum of the temperature fluctuations in the atmosphere by Champagne *et al* [9] and by Williams and Paulson [10] showed that at high wave numbers near $1/l_o$, the temperature spectrum features a small rise (or bump) that causes the spectrum to decrease less rapidly than predicted by the $\kappa^{-11/3}$ law of Oboukhov [11] and Corrsin [12]. Since the refractive index spectrum nominally follows the same spectral law as that of the temperature, this characteristic bump must also feature in its spectrum. However, none of the above spectrum models show this feature. In this regard, starting with first principles, Hill developed a theoretical spectrum [13, 14] for temperature fluctuations in the atmosphere which is in good agreement with the experimentally obtained data. For optical refractive index fluctuations in the atmosphere, the Hill spectrum $\Phi_n^H(\kappa)$ is the solution to a second-order linear homogeneous differential equation [15]:

$$\frac{d}{d\kappa}\left\{\kappa^{14/3}[(13.9\kappa\eta)^{3.8} + 1]^{-0.175}\frac{d}{d\kappa}\Phi_n^H(\kappa)\right\} = 14.1\kappa^4\eta^{4/3}\Phi_n^H(\kappa), \tag{7.28}$$

where $\eta = 0.135 l_o$. Hill and Clifford showed that the use of Hill spectrum in optical propagation calculations (like the variance of log intensity, the structure function of phase, and phase coherence length) can produce significantly different results than the Tatarskii spectrum [14]. Churnside provided the following analytical approximation to the Hill spectrum [15]:

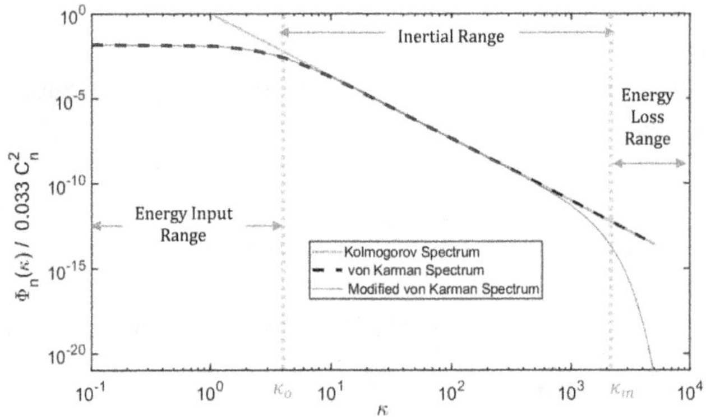

Figure 7.3. The Kolmogorov, von Kármán and modified von Kármán spectrum models for refractive index fluctuations are plotted with respect to wave number $\kappa(\text{m}^{-1})$ where $l_o = 5$ mm and $L_o = 2$ m.

$$\Phi_n^H(\kappa) \approx \Phi_n^K(\kappa)\{\exp(-70.5\kappa^2\eta^2) + 1.45\exp[-0.97(\ln(\kappa\eta) + 1.55)^2]\}, \quad (7.29)$$

where $\Phi_n^K(\kappa)$ is the Kolmogorov spectrum defined in equation (7.24). The above form of the Hill spectrum is useful for numerical integration of quantities that involve the refractive index power spectrum, but is not convenient for pure analytical studies. In this regard, Andrews gave the following approximate analytical form [16] of the Hill spectrum which is more convenient to use:

$$\Phi_n^M(\kappa) = 0.033C_n^2\left[1 + 1.802(\kappa/\kappa_l) - 0.254(\kappa/\kappa_l)^{7/6}\right]\frac{\exp(-\kappa^2/\kappa_l^2)}{(\kappa_o^2 + \kappa^2)^{11/6}}, \quad \text{for } 0 \leqslant \kappa < \infty \quad \kappa_l = 3.3/l_o \quad (7.30)$$

This approximation is also known as the modified atmospheric spectrum [3].

7.2.5 Behavior of turbulence strength: C_n^2 models

As seen from all the spectrum models discussed in the previous section, the refractive index structure constant C_n^2 plays a very important role in the description of the atmospheric turbulence. In the above description, we have assumed C_n^2 to be a constant, representing an averaged quantity. However, in reality C_n^2 is a random variable and depends on the temperature instabilities at a given geographical location and time. In general, C_n^2 is a seen to be a function of altitude, location and the time of the measurement [4]. Extensive experimental research has been going on to measure the C_n^2 profiles at various altitudes and locations resulting in the development of different mathematical models. The C_n^2 measurements have been collected at various locations and altitudes at different times of the day and year using optical scintillometers. These devices include a transmitter which emits electromagnetic radiation parallel to the propagation path and a receiver situated at some distance z from the transmitting aperture. The receiver measures the scintillation (or intensity fluctuations) in the beam due to the turbulence present in the atmospheric path. The numerical value of the intensity scintillation is then equated to the theoretical Rytov variance values and an estimate for C_n^2 is obtained. The generally observed values for the C_n^2 at a height of 2 m above ground is $10^{-17}\ \text{m}^{-2/3} < C_n^2 < 10^{-12}\ \text{m}^{-2/3}$ on a typical sunny day [5]. The important fluctuations in C_n^2 can be attributed to the following factors [7].

1. **Temperature fluctuations**: The C_n^2 values when measured over a 24-hour period show a characteristic cyclic pattern. It has been observed that the C_n^2 shows an increasing trend till late afternoon and then starts falling to lower values after twilight. Minimum values of C_n^2 are usually observed at sunrise and sunset.
2. **Altitude variation**: Many propagation problems of practical interest involve vertical or slanted beam propagation paths. In these cases, it becomes necessary to know the behavior of C_n^2 with altitude. The experimental studies to measure the C_n^2 profiles along a vertical path have been undertaken by mounting temperature sensors at various heights, for example, via use of

tower-, aircraft- or balloon-mounted sensors. The C_n^2 values are obtained by measuring the C_T^2 (temperature fluctuation) values at different heights [4]. It is observed that with increasing altitude h, the C_n^2 values approximately decrease as per the power law $h^{-4/3}$. Many C_n^2 distributions as a function of altitude have been proposed [7], e.g. the Hufnagel–Valley model, the Greenwood model and the Submarine Laser Communication Day (SLC-Day) model. The commonly used Hufnagel–Valley model approximates the C_n^2 profile as:

$$\begin{aligned} C_n^2(h) = {} & 5.94 \times 10^{-53}(v/27)^2 h^{10} \exp(-h/100) \\ & + 2.7 \times 10^{-16} \exp(-h/1500) + A \exp(-h/100), \end{aligned} \tag{7.31}$$

where, A sets the turbulence strength near ground with typical value of $A = 1.7 \times 10^{-14}\ \mathrm{m}^{-2/3}$. The parameter v is the high altitude wind speed with commonly used value for this parameter is $v = 21\ \mathrm{m\ s}^{-1}$.

Description of laser beam propagation in atmospheric turbulence thus needs careful attention to various models and parameters as discussed here. It is important to understand the approximations or empirical nature associated with these models when trying to simulate laser beam propagation through atmosphere at a specific geographical location. Empirical measurements of years'-long variations in quantities such as C_n^2, temperature, pressure, etc at specific locations may therefore be required before prediction of experimental behavior of laser beam profiles can be made satisfactorily using numerical simulations.

7.3 Classical perturbation methods

The classical perturbation techniques include Born approximation and Rytov approximation. Both of these perturbation methods are restricted to weak turbulence regimes. These approaches do take into account diffraction effects unlike the geometrical optics model. We will describe these historically important methods in some detail in the following discussion.

7.3.1 Born approximation

The first classical perturbation technique that we will discuss is Born's approximation. This technique was first used in quantum mechanics to study scattering problems. Light experiences scattering on interacting with the turbulent eddies present in the atmosphere. However, the term $n_1(\boldsymbol{r})$ which represents the fluctuations in the refractive index about its mean value is very small, in fact it is of the order of a few parts in 10^5 to 10^6. It is reasonable to expect that the light beam will propagate through the atmosphere very much like it does in free space but with additional small variations as a result of the scattering. In the weak scattering limit, Born's approximation states that the field $U(\boldsymbol{r})$ can be written as a sum of terms of the form:

$$U(\boldsymbol{r}) = U_0(\boldsymbol{r}) + \nu U_1(\boldsymbol{r}) + \nu^2 U_2(\boldsymbol{r}) + \cdots \tag{7.32}$$

where ν is a dummy variable that shows the order of smallness of the terms in the expansion. The first term $U_0(\boldsymbol{r})$ represents the unscattered portion of the field or, in other words, the field which would be obtained on free-space propagation. The successive terms represent the scattered portion of the field, for example, $U_1(\boldsymbol{r})$ corresponds to single scattering of the incident field by a turbulent eddy and $U_2(\boldsymbol{r})$ represents double scattering and so on. If magnitude of the each successive term is sufficiently smaller than its immediate preceding term, i.e., $|U_{m+1}| \ll |U_m|$, then the series will converge and a finite number of terms will be able to completely describe the solution. However, this may not be true for all propagation problems, specifically when the total integrated refractive index fluctuations over the propagation path become large. This can happen due to either an increase in $n_1(\boldsymbol{r})$ or an increase in the length of the propagation path. In such cases, multiple scattering effects become dominant and the basic condition of weak scattering as required in Born's approximation is violated (figure 7.4). The average value of refractive index in the atmosphere is $\langle n(\boldsymbol{r})\rangle \approx 1$, so using equation (7.6), we can write

$$n^2(\boldsymbol{r}) = (1 + \nu n_1(\boldsymbol{r}))^2 \approx 1 + 2\nu n_1(\boldsymbol{r}) \tag{7.33}$$

where we have neglected the second-order term $n_1^2(\boldsymbol{r})$ as it is very small in magnitude. Substituting for $U(\boldsymbol{r})$ and $n^2(\boldsymbol{r})$ in equation (7.8) from equations (7.32)

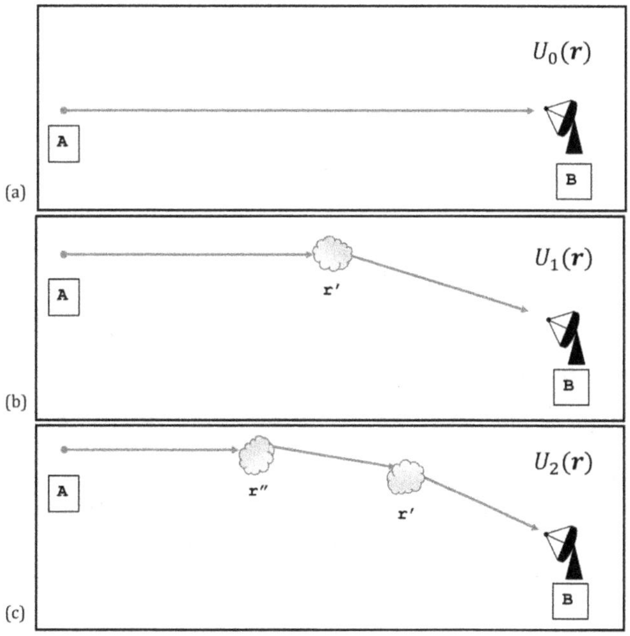

Figure 7.4. Schematic diagram showing the first three terms in Born's approximation. The transmitter at point A emits a plane wave which travels through the medium to the receiver situated at point B. The first term in Born's approximation $U_0(\boldsymbol{r})$ represents the unperturbed wave as shown in (a). The second term $U_1(\boldsymbol{r})$ and the third term $U_2(\boldsymbol{r})$ represent the single and double scattered radiation, respectively. These cases are represented in (b) and (c), respectively.

and (7.33), respectively, and equating terms corresponding to equal powers of ν to zero yields the following system of equations:

$$\nu^0\text{: } \nabla^2 U_0(\boldsymbol{r}) + k^2 U_0(\boldsymbol{r}) = 0 \tag{7.34}$$

$$\nu^1\text{: } \nabla^2 U_1(\boldsymbol{r}) + k^2 U_1(\boldsymbol{r}) = -2k^2 n_1(\boldsymbol{r}) U_0(\boldsymbol{r}) \tag{7.35}$$

$$\nu^2\text{: } \nabla^2 U_2(\boldsymbol{r}) + k^2 U_2(\boldsymbol{r}) = -2k^2 n_1(\boldsymbol{r}) U_1(\boldsymbol{r}) \tag{7.36}$$

and so on for higher values of ν. An important outcome of Born's approximation is that we have broken down the original stochastic Helmholtz equation which was a homogeneous partial differential equation with random coefficients into a system of nonhomogeneous partial differential equations with constant coefficients and source terms. These equations can be solved using Green's function method. The partial differential equation (equation (7.35)) corresponding to the first-order Born's approximation term $U_1(\boldsymbol{r})$ is usually solved. Therefore, Born's approximation is also sometimes called the *single scattering approximation* as we are looking for solutions of the form $U_B^{(1)}(\boldsymbol{r}) = U_0(\boldsymbol{r}) + U_1(\boldsymbol{r})$, neglecting all higher order scattering terms The solution of equation (7.35) is equal to the convolution of the free-space Green's function $G(\boldsymbol{r})$ with the source term $-2k^2 n_1(\boldsymbol{r}) U_0(\boldsymbol{r})$. Green's function in free space is given by:

$$G(\boldsymbol{r}) = \frac{1}{4\pi} \frac{e^{ik|\boldsymbol{r}|}}{|\boldsymbol{r}|} \tag{7.37}$$

Then, the solution for $U_1(\boldsymbol{r})$ can be written as:

$$U_1(\boldsymbol{r}) = \frac{1}{4\pi} \int_V \frac{e^{ik|\boldsymbol{r}-\boldsymbol{r}'|}}{|\boldsymbol{r}-\boldsymbol{r}'|} [2k^2 n_1(\boldsymbol{r}') U_0(\boldsymbol{r}')] d^3\boldsymbol{r}' \tag{7.38}$$

where V is the scattering volume. At any position $\boldsymbol{r}'$ inside the scattering volume V, the incident field $U_0(\boldsymbol{r}')$ interacts with the turbulent eddies present in the atmosphere and gets scattered in the form of secondary spherical waves which are collected at the receiver situated at $\boldsymbol{r}$. Hence, the first correction term of the Born's approximation $U_1(\boldsymbol{r})$ is just the summation of these scattered spherical waves inside the scattering volume where the strength of these waves is given by the product of the incident field $U_0(\boldsymbol{r}')$ and the refractive index fluctuations $n_1(\boldsymbol{r}')$. An important point to note is that by definition, it is assumed that $\langle n_1(\boldsymbol{r}) \rangle = 0$, then equation (7.38), implies that the ensemble average or mean of the first Born approximation term also vanishes $\langle U_1(\boldsymbol{r}) \rangle = 0$. Higher order perturbation terms $U_2(\boldsymbol{r})$, $U_3(\boldsymbol{r})$, etc can be solved in the same manner, however, we will not discuss these solutions here.

7.3.1.1 Paraxial approximation

The solution for $U_1(\boldsymbol{r})$ as given by equation (7.38) can be converted into a much simpler form by making use of certain approximations. We begin by assuming that the backscattering of the incident wave can be neglected completely. This is a direct consequence of the weak scattering approximation where the incident wave is assumed to be scattered only in the forward direction by the weak turbulent eddies. The smallest sized eddies are responsible for the largest angle of scattering. Therefore, considering the inner scale of turbulence to be equal to l_o, the maximum scattering angle would be of the order of $\theta = \lambda/l_o$. The smallest eddies in the atmosphere are of the order of a few millimeters, so that $\lambda \ll l_o$. The scattering angle for typical values of $\lambda = 0.6\,\mu\text{m}$ and $l_o = 2$ mm comes out to be equal to $\theta = 3 \times 10^{-4}$ radians, which is small enough. Figure 7.5 shows the scattering of an incident plane wave by turbulent eddies of various spatial sizes l. In order for the scattered waves from any turbulent eddy to reach the receiver situated at P, the maximum lateral displacement of that turbulent eddy from the propagation axis should follow the relation:

$$\rho \leqslant \frac{\lambda z}{l_o} \tag{7.39}$$

This condition limits the lateral extent of the turbulent eddies which contribute to the optical field received at P. In other words, this means that the turbulent eddies which are far off from the propagation axis result in scattered waves which never reach the receiver. Therefore, the received optical field at P can be visualized to contain waves scattered in a narrow cone about the wave propagation direction. Typically, the longitudinal distance z from the scatterer to receiver is much larger than the transverse displacement ρ from the z-axis, i.e., $z \gg \rho$. This allows us to use *paraxial approximation* to expand the exponential term in the Green's function of equation (7.37) in a binomial series:

$$k|\boldsymbol{r}| = kz\sqrt{1 + \left(\frac{\rho}{z}\right)^2} = kz\left[1 + \frac{1}{2}\left(\frac{\rho}{z}\right)^2 - \frac{1}{8}\left(\frac{\rho}{z}\right)^4 + \cdots\right] \tag{7.40}$$

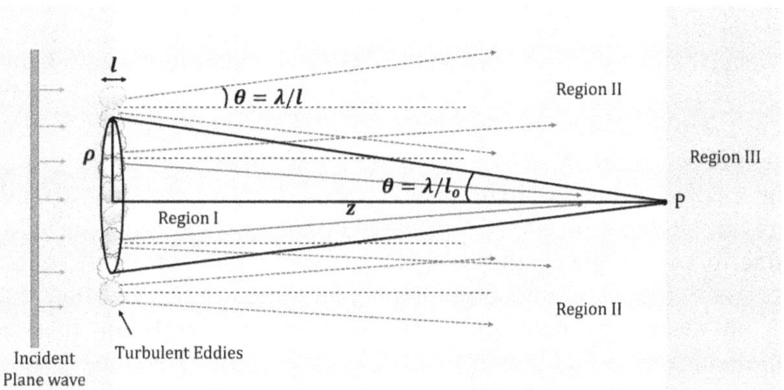

Figure 7.5. The figure shows the forward scattering of an incident plane wave by the turbulent eddies of spatial sizes 'l'. The eddies scatter the incident radiation at narrow angles given by $\theta = \lambda/l$. The major contribution to the optical field at point P is from the radiation scattered from eddies situated near to the z-axis, as represented by region I. The contribution from the forward-scattered radiation from region II to the field at point P is very small. In region III, only backscattered radiation reaches P.

The first term in this expansion gives the phase evolution of a wave in free space and is always retained in the expansion. The higher order terms are retained if they are significant enough to cause a phase change of more than 2π. The second term is of the order of $\pi\rho^2/\lambda z$ where the maximum value of ρ is set by equation (7.39) and $\sqrt{\lambda z}$ is called the *Fresnel length* of the system. For a typical propagation distance of $z = 1$ km, $\lambda = 0.6\ \mu$m and $l_o = 2$ mm, the second term remains large and is retained in the expansion. However, the third term is of the order of $\pi\rho^4/4\lambda z^3$ whose value is much much less than 1 radian for the above parameters and is therefore inconsequential to us and can be safely dropped. Similarly, the denominator in the Green's function can be simply approximated as $|\boldsymbol{r}| \approx z$. This allows us to write solution $U_1(\boldsymbol{r})$ in a simple form:

$$U_1(\boldsymbol{r}) = \frac{k^2}{2\pi}\int_V \exp\left(ik\left[(z - z') + \frac{(\rho - \rho')^2}{2(z - z')}\right]\right)\frac{n_1(\boldsymbol{r}')U_0(\boldsymbol{r}')}{(z - z')}d^3\boldsymbol{r}' \qquad (7.41)$$

The volume integration in the above equation can be separated into three regions, as shown in figure 7.5. Region I is the weakly scattered waves from the turbulent eddies contained in a narrow cone of angle $\theta = \lambda/l_o$ as explained earlier. It represents the most significant contribution to the optical field at point *P*. Due to the condition that wavelength λ is much smaller compared to the inner scale l_o of turbulence, the contribution from regions II and III which represents large-angle scattering and backscattering, respectively, is negligible. The contribution from region II and region III to the integral in equation (7.41) can be shown to be small (of the order of $\approx n_1^2/16$) using the method of stationary phase. Large-angle scattering is also not of utmost importance in understanding the laser beam profile after propagating a distance of a few km from the laser source through turbulence. The light received by a detector or receiver at such large distances mainly represents small spatial frequencies associated with scattered light.

7.3.1.2 Depolarization of light

Let us revisit equation (7.1) and pay some attention to the depolarization term which we had ignored all this while. The scattering in the atmosphere can give rise to subtle changes in the polarization of the light beam. As discussed earlier, for long-range line of sight propagation the scattering angles involved are usually very small and so significant changes in the polarization solely due to the refractive index inhomogeneities are not observed. In such cases, the major depolarization of the light beam is caused due to wide-angle scattering from aerosols, pollutants, dust particles, rain drops etc present in the atmosphere. However, there might be some wide-angle scattering in the troposphere which can give rise to significant changes in the polarization of the beam for propagation beyond the horizon. Two different approaches based on diffraction theory and geometrical optics have been developed for describing the depolarization of light by the atmosphere. Here, we would look into the calculation first provided by Tatarskii using Born's approximation [17].

Let us assume that a linearly polarized plane wave is traveling along the z-axis and we want to calculate by how much its electric field vector gets rotated on propagation through the atmosphere. Assuming a sinusoidal time dependence for the electric field (see equation (7.2)), we can write equation (7.1) as:

$$\nabla^2 \boldsymbol{E}_o(\boldsymbol{r}) + \boldsymbol{\nabla}(\boldsymbol{E}_o(\boldsymbol{r}).\ \boldsymbol{\nabla} \ln \varepsilon) + k^2 n^2(\boldsymbol{r}) \boldsymbol{E}_o(\boldsymbol{r}) = 0 \tag{7.42}$$

Using Born's approximation, we write $\boldsymbol{E}_o(\boldsymbol{r})$ as:

$$\boldsymbol{E}_o(\boldsymbol{r}) = \boldsymbol{E}_0(\boldsymbol{r}) + \nu \boldsymbol{E}_1(\boldsymbol{r}) \tag{7.43}$$

where $\boldsymbol{E}_0(\boldsymbol{r})$ is the incident field and $\boldsymbol{E}_1(\boldsymbol{r})$ represents the scattered field. Let the incident field be x-polarized, then $\boldsymbol{E}_0(\boldsymbol{r}) = E_x^0(\boldsymbol{r})\hat{\boldsymbol{x}}$ but the scattered field also contains y-polarized components due to depolarization, therefore $\boldsymbol{E}_1(\boldsymbol{r}) = E_x^1(\boldsymbol{r})\hat{\boldsymbol{x}} + E_y^1(\boldsymbol{r})\hat{\boldsymbol{y}}$. Substituting for the quantities $\boldsymbol{E}_o(\boldsymbol{r})$, $n^2(\boldsymbol{r})$ and $\ln \varepsilon = 2 \ln n$ in equation (7.42) to collect equal powers of ν:

$$\nu^0\!: \ \nabla^2 \boldsymbol{E}_0(\boldsymbol{r}) + k^2 \boldsymbol{E}_0(\boldsymbol{r}) = 0 \tag{7.44}$$

$$\nu^1\!: \ \nabla^2 \boldsymbol{E}_1(\boldsymbol{r}) + k^2 \boldsymbol{E}_1(\boldsymbol{r}) = -2k^2 n_1(\boldsymbol{r}) \boldsymbol{E}_0(\boldsymbol{r}) - 2\nabla(\boldsymbol{E}_0(\boldsymbol{r}).\ \nabla n_1(\boldsymbol{r})) \tag{7.45}$$

Here, $\ln(1 + n_1(r))$ has been expanded in a power series in $n_1(r)$. The depolarized component of the scattered wave is given by:

$$\nabla^2 E_y^1 + k^2 E_y^1 = -2E_x^0(z) \frac{\partial^2 n_1}{\partial y \partial x} \tag{7.46}$$

Therefore, the solution of the depolarized component can be obtained using Green's function and is given by:

$$E_y^1(\boldsymbol{r}) = \frac{1}{4\pi} \int_V \frac{e^{ik|\boldsymbol{r}-\boldsymbol{r}'|}}{|\boldsymbol{r} - \boldsymbol{r}'|} [2E_x^0(z') \frac{\partial^2 n_1(\boldsymbol{r}')}{\partial y' \partial x'}] d^3\boldsymbol{r}' \tag{7.47}$$

The mean-squared polarization fluctuation is defined as the average intensity of the depolarized component divided by the intensity of the incident plane wave:

$$\langle M^2 \rangle = \left\langle \left| \frac{E_y^1(\boldsymbol{r})}{E_x^0(\boldsymbol{r})} \right|^2 \right\rangle \tag{7.48}$$

Substituting for $E_y^1(\boldsymbol{r})$ in this expression, we can write:

$$\langle M^2 \rangle = \frac{1}{4\pi^2} \int_V d^3\boldsymbol{r}' \left(\frac{e^{ik|\boldsymbol{r}-\boldsymbol{r}'|}}{|\boldsymbol{r} - \boldsymbol{r}'|} \frac{E_x^0(z')}{E_x^0(\boldsymbol{r})} \right) \int_V d^3\boldsymbol{r}'' \left(\frac{e^{ik|\boldsymbol{r}-\boldsymbol{r}''|}}{|\boldsymbol{r} - \boldsymbol{r}''|} \frac{E_x^0(z'')}{E_x^0(\boldsymbol{r}'')} \right)^* \left\langle \frac{\partial^2 n_1(\boldsymbol{r}')}{\partial y' \partial x'} \frac{\partial^2 n_1(\boldsymbol{r}'')}{\partial y'' \partial x''} \right\rangle \tag{7.49}$$

The quantity in the angle brackets which gives the ensemble average of the double gradient product can be written in terms of the refractive index spectrum [17, 18]. When the refractive index inhomogeneities are assumed to follow the Tatarskii spectrum with an inner scale cutoff wave number κ_m (see equation (7.25)), then for the case of a point receiver, $\langle M^2 \rangle$ is given by [17, 18]:

$$\langle M^2 \rangle = 0.070 C_n^2 \kappa_m^{7/3} k^{-2} L \qquad (7.50)$$

where L denotes the total path length. Assuming $l_o = 2$ mm, $\lambda = 0.6\ \mu$m, $L = 1$ km and $C_n^2 = 10^{-13}$, we get $\langle M^2 \rangle = 8 \times 10^{-18}$ which is way too small to be detected by the conventional optical systems [19, 20].

7.3.2 Rytov approximation

In the stochastic Helmholtz equation (equation (7.8)) the second term is a product of the stochastic function $n(\boldsymbol{r})$ with the scalar field $U(\boldsymbol{r})$. This makes it very difficult to solve this equation as the coefficients of the required solution $U(\boldsymbol{r})$ are random functions of space. Rytov transformation provides a means to separate these two functions and convert this equation into the *Riccati equation.* In this newer form, the randomly varying function $n(\boldsymbol{r})$ appears as an additional source term. This transformation was first described by Rytov in 1937, during his investigations on diffraction of light by ultrasonic waves [21]. Later on, Obukhov used this transformation to study wave propagation in a random medium [22]. Rytov's transformation involves substituting the scalar field $U(\boldsymbol{r})$ as

$$U(\boldsymbol{r}) = U_0(\boldsymbol{r})\exp(\psi(\boldsymbol{r})) \qquad (7.51)$$

into the Helmholtz equation where $U_0(\boldsymbol{r})$ represents the unperturbed field in the absence of turbulence. Here, the function $\psi(\boldsymbol{r}) = \chi + iS$ is a complex quantity which represents the effect of the turbulence on the beam. This substitution leads to an equation of the form:

$$\nabla^2 U_0(\boldsymbol{r}) + 2[\nabla U_0(\boldsymbol{r}).\ \nabla\psi(\boldsymbol{r})] + U_0(\boldsymbol{r})[\nabla^2\psi(\boldsymbol{r}) + (\nabla\psi(\boldsymbol{r}))^2] + k^2(1 + 2n_1(\boldsymbol{r}))U_0 = 0 \quad (7.52)$$

As the field $U_0(\boldsymbol{r})$ is the free-space field therefore, it satisfies the homogeneous wave equation and we have

$$\nabla^2 U_0(\boldsymbol{r}) + k^2 U_0(\boldsymbol{r}) = 0 \qquad (7.53)$$

which can be substituted in the above equation. Let us now assume that the field $U_0(\boldsymbol{r})$ can be further expressed as an exponential of a function $\psi_0(\boldsymbol{r})$ such that, $U_0(\boldsymbol{r}) = \exp(\psi_0(\boldsymbol{r}))$, then the total field $U(\boldsymbol{r})$ is simply given by

$$U(\boldsymbol{r}) = \exp(\psi_0(\boldsymbol{r}) + \psi(\boldsymbol{r})) \qquad (7.54)$$

Therefore, the complex function $\psi(\boldsymbol{r})$ satisfies the equation:

$$\nabla^2\psi(\boldsymbol{r}) + [\nabla\psi(\boldsymbol{r})]^2 + 2[\nabla\psi_0.\ \nabla\psi(\boldsymbol{r})] + 2k^2 n_1(\boldsymbol{r}) = 0 \qquad (7.55)$$

It can be observed that the required solution $\psi(\boldsymbol{r})$ of this non-linear partial differential equation has been separated from the stochastic function $n_1(\boldsymbol{r})$.

Tatarskii made extensive use of the Rytov transformation in his seminal works on wave propagation through turbulence [2, 23] and described a perturbative technique known as the *Rytov approximation* or *method of smooth perturbations* for such problems. In the weak scattering limit, the phase function $\psi(\boldsymbol{r})$ is written as a sum of terms of form:

$$\psi(\boldsymbol{r}) = \nu\psi_1(\boldsymbol{r}) + \nu^2\psi_2(\boldsymbol{r}) + \cdots \tag{7.56}$$

where ν is a dummy variable that shows the order of smallness of the terms in the expansion. If we compare the current expression for $U(\boldsymbol{r})$ with Born's approximation expression (given by equation(7.32)), we notice a major difference between the two treatments. In Born's approximation, the scalar field $U(\boldsymbol{r})$ is represented as an additive perturbation series where the field contributions due to the scattering are added to the initial unperturbed wave field. However, the Rytov approximation involves a multiplicative perturbation series.

We can substitute the expansion for $\psi(\boldsymbol{r})$ and $n^2(\boldsymbol{r})$ from equation(7.56) and equation(7.33) into the above non-linear equation (7.55). The terms containing equal powers of ν can be collected to write the following equations:

$$\nu^0\text{:}\ \ \nabla^2\psi_0(\boldsymbol{r}) + [\nabla\psi_0(\boldsymbol{r})]^2 + k^2 = 0 \tag{7.57}$$

$$\nu^1\text{:}\ \ \nabla^2\psi_1(\boldsymbol{r}) + 2[\nabla\psi_0(\boldsymbol{r}).\ \ \nabla\psi_1(\boldsymbol{r})] = -2k^2 n_1(\boldsymbol{r}) \tag{7.58}$$

$$\nu^2\text{:}\ \ \nabla^2\psi_2(\boldsymbol{r}) + 2[\nabla\psi_0(\boldsymbol{r}).\ \ \nabla\psi_2(\boldsymbol{r})] = -\nabla\psi_1(\boldsymbol{r}).\ \ \nabla\psi_1(\boldsymbol{r}) \tag{7.59}$$

The equation corresponding to ν^0 gives the free-space propagation of the wave. The equations corresponding to higher powers of $\nu > 1$ all have a general form:

$$\nabla^2\psi_\nu(\boldsymbol{r}) + 2[\nabla\psi_0(\boldsymbol{r}).\ \ \nabla\psi_\nu(\boldsymbol{r})] = -\sum_{j=1}^{\nu-1}\nabla\psi_j(\boldsymbol{r}).\ \ \nabla\psi_{\nu-j}(\boldsymbol{r}) = f_\nu(\boldsymbol{r}) \tag{7.60}$$

The usual practice is to solve only for the first-order term. This solution is known as the basic Rytov solution and is used widely to describe beam propagation through a random medium. The equations for the higher order terms can be solved if $f_\nu(\boldsymbol{r})$ is known, which itself depends on the lower-order solutions. The series for $\psi(\boldsymbol{r})$ would converge if the successive terms ψ_{m+1} are smaller than their preceding terms ψ_m.

The equation for the first-order term $\psi_1(\boldsymbol{r})$ depends only on the function $\psi_0(\boldsymbol{r})$ and the refractive index variation $n_1(\boldsymbol{r})$. Let us assume the function $\psi_1(\boldsymbol{r})$ to be in the form:

$$\psi_1(\boldsymbol{r}) = Q(\boldsymbol{r})\exp(-\psi_0(\boldsymbol{r})) \tag{7.61}$$

This simplifies the equation (7.58) to:

$$\nabla^2 Q(\boldsymbol{r}) - Q[\nabla^2\psi_0(\boldsymbol{r}) + (\nabla\psi_0(\boldsymbol{r}))^2] = -2k^2 n_1(\boldsymbol{r})e^{\psi_0(\boldsymbol{r})} \tag{7.62}$$

The term inside the brackets on the left-hand side of this equation can be substituted from the equation (7.57) and we get:

$$\nabla^2 Q(\boldsymbol{r}) + k^2 Q(\boldsymbol{r}) = -2k^2 n_1(\boldsymbol{r})e^{\psi_0(\boldsymbol{r})} \tag{7.63}$$

This equation is solvable using the method of Green's function as explained in the previous section. Therefore, we can write the solution for $W(\boldsymbol{r})$ as:

$$Q(\boldsymbol{r}) = \frac{1}{4\pi}\int_V \frac{e^{ik|\boldsymbol{r}-\boldsymbol{r}'|}}{|\boldsymbol{r}-\boldsymbol{r}'|}[2k^2 n_1(\boldsymbol{r}')e^{\psi_0(\boldsymbol{r}')}]d^3\boldsymbol{r}' \tag{7.64}$$

The basic Rytov solution is obtained as:

$$\psi_1(\boldsymbol{r}) = \frac{1}{4\pi}\int_V \frac{e^{ik|\boldsymbol{r}-\boldsymbol{r}'|}}{|\boldsymbol{r}-\boldsymbol{r}'|}[2k^2 n_1(\boldsymbol{r}')e^{\psi_0(\boldsymbol{r}')-\psi_0(\boldsymbol{r})}]d^3\boldsymbol{r}' \tag{7.65}$$

or using the relation $U_0(\boldsymbol{r}) = \exp(\psi_0(\boldsymbol{r}))$, one can write

$$\psi_1(\boldsymbol{r}) = \frac{1}{4\pi}\int_V \frac{e^{ik|\boldsymbol{r}-\boldsymbol{r}'|}}{|\boldsymbol{r}-\boldsymbol{r}'|}\left[2k^2 n_1(\boldsymbol{r}')\frac{U_0(\boldsymbol{r}')}{U_0(\boldsymbol{r})}\right]d^3\boldsymbol{r}' \tag{7.66}$$

On comparing this relation with equation (7.38), we can see that

$$\psi_1(\boldsymbol{r}) = \frac{U_1(\boldsymbol{r})}{U_0(\boldsymbol{r})} \tag{7.67}$$

where $U_1(\boldsymbol{r})$ is the first-order scattering term of Born's approximation. The basic Rytov solution can thus be written as:

$$U_R^{(1)}(\boldsymbol{r}) = U_0(\boldsymbol{r})\exp(\psi_1(\boldsymbol{r})) = U_0(\boldsymbol{r})\exp[U_1(\boldsymbol{r})/U_0(\boldsymbol{r})] \tag{7.68}$$

In the limit of weak scattering where $U_1(\boldsymbol{r}) \ll U_0(\boldsymbol{r})$, the exponential in the above equation can be expanded in a series, which is equal to the first-order Born approximation solution:

$$U_R^{(1)}(\boldsymbol{r}) = U_0(\boldsymbol{r})\left[1 + \frac{U_1(\boldsymbol{r})}{U_0(\boldsymbol{r})}\right] = U_0(\boldsymbol{r}) + U_1(\boldsymbol{r}) = U_B^{(1)}(\boldsymbol{r}) \tag{7.69}$$

It was further shown by Yura [24, 25] that in the weak scattering limit, the solution of the second-order Rytov approximation term $\psi_2(\boldsymbol{r})$, can be written as a combination of the first- and second-order Born's approximation terms:

$$\psi_2(\boldsymbol{r}) = \frac{U_2(\boldsymbol{r})}{U_0(\boldsymbol{r})} - \frac{1}{2}\left[\frac{U_1(\boldsymbol{r})}{U_0(\boldsymbol{r})}\right]^2 \tag{7.70}$$

This suggests an interesting relationship between the two perturbation methods. Let us try to understand this more by looking at the fourth-order solutions obtained from both Rytov and Born methods:

$$U_R^{(4)} = U_0(\boldsymbol{r})\exp[\nu\psi_1(\boldsymbol{r}) + \nu^2\psi_2(\boldsymbol{r}) + \nu^3\psi_3(\boldsymbol{r}) + \nu^4\psi_4(\boldsymbol{r})] \tag{7.71}$$

$$U_B^{(4)} = U_0(\boldsymbol{r})[1 + \nu\Phi_1(\boldsymbol{r}) + \nu^2\Phi_2(\boldsymbol{r}) + \nu^3\Phi_3(\boldsymbol{r}) + \nu^4\Phi_4(\boldsymbol{r})] \tag{7.72}$$

where $\Phi_m = U_m(\boldsymbol{r})/U_0(\boldsymbol{r})$ gives the *normalized Born perturbation terms* and ν is a dummy variable which gives the order of smallness of the terms. On equating the two solutions we get:

$$\nu\psi_1(\boldsymbol{r}) + \nu^2\psi_2(\boldsymbol{r}) + \nu^3\psi_3(\boldsymbol{r}) + \nu^4\psi_4(\boldsymbol{r}) = \ln[1 + \nu\Phi_1(\boldsymbol{r}) + \nu^2\Phi_2(\boldsymbol{r}) + \nu^3\Phi_3(\boldsymbol{r}) + \nu^4\Phi_4(\boldsymbol{r})] \tag{7.73}$$

As the magnitude of the Born approximation terms $U_m(\boldsymbol{r})$ are assumed to be smaller than the unscattered term $U_0(\boldsymbol{r})$, therefore the natural logarithm in equation (7.73) can be expanded in a Maclaurin series using the relation:

$$\ln(1 + x) = x - \frac{x^2}{2} + \frac{x^3}{3} - \frac{x^4}{4} + \cdots \tag{7.74}$$

By equating the equal powers of ν on both sides of the equation, the first four Rytov perturbation terms can be approximated as:

$$\psi_1(\boldsymbol{r}) = \Phi_1(\boldsymbol{r}) \tag{7.75}$$

$$\psi_2(\boldsymbol{r}) = \Phi_2(\boldsymbol{r}) - \frac{1}{2}\Phi_1^2(\boldsymbol{r}) \tag{7.76}$$

$$\psi_3(\boldsymbol{r}) = \Phi_3(\boldsymbol{r}) - \Phi_1(\boldsymbol{r})\Phi_2(\boldsymbol{r}) + \frac{1}{3}\Phi_1^3(\boldsymbol{r}) \tag{7.77}$$

$$\psi_4(\boldsymbol{r}) = \Phi_4(\boldsymbol{r}) - \frac{1}{2}\Phi_2^2(\boldsymbol{r}) - \frac{1}{4}\Phi_1^4(\boldsymbol{r}) - \Phi_1(\boldsymbol{r})\Phi_3(\boldsymbol{r}) + \Phi_1^2(\boldsymbol{r})\Phi_2(\boldsymbol{r}) \tag{7.78}$$

This analysis suggests that the Rytov perturbation terms can be approximately expressed by a sum of different perturbation terms of the Born series. However, there are some important differences between the two treatments which become apparent when the phase fluctuations in the beam are large. In such cases, the exponential term in Rytov method cannot be approximated by Born perturbation terms as now $U_m(\boldsymbol{r})$ might not be very small compared to the unscattered term $U_0(\boldsymbol{r})$. In such cases, the Rytov method provides a superior description and is preferred over the Born series.

7.3.2.1 Amplitude and phase fluctuations

We have established the perturbative solutions using both Born and Rytov approximation methods. We are now in a position to investigate amplitude and phase fluctuations experienced by the incident optical field on its interaction with the turbulent atmosphere. Let us begin by assuming that the field after experiencing turbulence $U(\boldsymbol{r})$ is represented by amplitude A and phase s while the incident (or the free-space) field $U_0(\boldsymbol{r})$ has amplitude A_0 and phase s_0, that is:

$$U(\boldsymbol{r}) = A\exp(is) \tag{7.79}$$

$$U_0(\boldsymbol{r}) = A_0\exp(is_0) \tag{7.80}$$

Then,

$$\frac{U(\boldsymbol{r})}{U_0(\boldsymbol{r})} = \frac{A}{A_0}\exp[i(s - s_0)] \tag{7.81}$$

Let $U(\boldsymbol{r})$ be given by the basic Rytov solution (see equation (7.68)), then one can write:

$$\frac{U(\boldsymbol{r})}{U_0(\boldsymbol{r})} = \exp(\psi_1) \tag{7.82}$$

Equating the above two equations and taking the logarithm on both sides gives:

$$\psi_1 = \ln\left[\frac{A}{A_0}\right] + i(s - s_0) \tag{7.83}$$

The log-amplitude fluctuations χ and the phase fluctuation S are defined as:

$$\chi = \ln\left[\frac{A}{A_0}\right] \tag{7.84}$$

$$S = s - s_0 \tag{7.85}$$

Therefore, one can write

$$\psi_1 = \chi + iS \tag{7.86}$$

Thus we see that the log-amplitude fluctuations and phase fluctuations are equal to the real and imaginary part of $\psi_1(\boldsymbol{r})$. By invoking the paraxial approximation described in the last section, one can write χ and S as:

$$\begin{bmatrix}\chi(\boldsymbol{r})\\ S(\boldsymbol{r})\end{bmatrix} = \begin{bmatrix}Re\\ Im\end{bmatrix}\left\{\frac{k^2}{2\pi}\int_V \exp\left(ik\left[(z - z') + \frac{(\rho - \rho')^2}{2(z - z')}\right]\right)\frac{n_1(\boldsymbol{r}')}{(z - z')}\frac{U_0(\boldsymbol{r}')}{U_0(\boldsymbol{r})}d^3\boldsymbol{r}'\right\} \tag{7.87}$$

For the present discussion, we would consider a unit amplitude plane wave propagating in $+z$ direction as the original incident (or unperturbed) field, i.e., $U_0(\boldsymbol{r}) = \exp(ikz)$. This reduces the above equation to the form:

$$\begin{bmatrix}\chi(\boldsymbol{r})\\ S(\boldsymbol{r})\end{bmatrix} = \begin{bmatrix}Re\\ Im\end{bmatrix}\left\{\frac{k^2}{2\pi}\int_V \exp\left[\frac{ik(\rho - \rho')^2}{2(z - z')}\right]\frac{n_1(\boldsymbol{r}')}{(z - z')}d^3\boldsymbol{r}'\right\} \tag{7.88}$$

The expressions for log-amplitude and phase fluctuations for the other forms of incident fields (spherical, Gaussian) can be found in a similar fashion. Also, note that though we are considering log-amplitude fluctuation, in the weak scatter limit, this quantity is identical to the normalized amplitude fluctuation: $\chi = \ln\left[\frac{A}{A_0}\right] \simeq \frac{A}{A_0}$.

The amplitude and phase fluctuation in the beam increases as the beam propagates along the z-axis and experiences more turbulence. For description of this random process, one needs to consider the averaged or statistical quantities. From previous discussion, we know that $\langle n_1(\boldsymbol{r})\rangle = 0$ and so the first-order moments for the log-amplitude and phase fluctuations also vanish, that is $\langle\chi(\boldsymbol{r})\rangle = \langle S(\boldsymbol{r})\rangle = 0$. This is true because in the current treatment, only the first-order perturbation term $\psi_1(\boldsymbol{r})$ has been used for calculation of the total field $U(\boldsymbol{r})$. If higher order perturbation terms are also included, then $\langle\chi(\boldsymbol{r})\rangle \neq \langle S(\boldsymbol{r})\rangle \neq 0$ but still these mean values would be inconsequentially small. In order to calculate the higher order statistical moments, it is useful to derive spectral representations of the amplitude and phase fluctuations first.

7.3.2.2 Spectral representation of amplitude and phase fluctuations
Consider the log-amplitude and phase fluctuations as given in last section:

$$\begin{bmatrix} \chi(\boldsymbol{r}) \\ S(\boldsymbol{r}) \end{bmatrix} = \frac{k^2}{2\pi}\int_V \frac{n_1(\boldsymbol{r}')}{(z - z')}\begin{bmatrix} \cos \\ \sin \end{bmatrix}\left[\frac{k(\rho - \rho')^2}{2(z - z')}\right] d^3\boldsymbol{r}' \tag{7.89}$$

The refractive index function $n_1(\boldsymbol{r}')$ can be expressed as a two-dimensional Fourier–Stieltjes integral:

$$n_1(\boldsymbol{r}') = \int d\nu(\boldsymbol{\kappa}, z')\exp(i\boldsymbol{\kappa}\cdot\boldsymbol{\rho}') \tag{7.90}$$

where $\boldsymbol{\kappa}$ represents the three-dimensional wave vector with components $(\kappa_x, \kappa_y, \kappa_z = 0)$. The function $\nu(\boldsymbol{\kappa}, z')$ is a random function of propagation distance z and it represents the amplitude of the refractive index fluctuations in a transverse plane perpendicular to the z-axis. On substituting this form of $n_1(\boldsymbol{r}')$ into equation(7.89) and expanding $d^3\boldsymbol{r}' = d^2\rho' dz'$, we can solve for $\chi(\boldsymbol{r})$[2],

$$\chi(\boldsymbol{r}) = \frac{k^2}{2\pi}\int\int_0^z dz'\frac{d\nu(\boldsymbol{\kappa}, z')}{(z - z')}\int d^2\rho' \exp(i\boldsymbol{\kappa}\cdot\boldsymbol{\rho}')\cos\left[\frac{k(\rho - \rho')^2}{2(z - z')}\right] \tag{7.91}$$

Let the integral over $\boldsymbol{\rho}'$ be denoted by I, this integral can be solved by substituting $\boldsymbol{\rho}' = \boldsymbol{\rho}'' + \boldsymbol{\rho}$, therefore one can write

$$I = \exp(i\boldsymbol{\kappa}\cdot\boldsymbol{\rho})\int d^2\rho'' \exp(i\boldsymbol{\kappa}\cdot\boldsymbol{\rho}'')\cos\left[\frac{k(\rho'')^2}{2(z - z')}\right] \tag{7.92}$$

$$I = \exp(i\boldsymbol{\kappa}\cdot\boldsymbol{\rho})\int_0^\infty \rho'' d\rho'' \cos\left[\frac{k(\rho'')^2}{2(z - z')}\right]\int_0^{2\pi} d\theta \exp\left[iK\rho''\cos(\theta - \alpha)\right] \tag{7.93}$$

where the polar coordinates $\boldsymbol{\rho}'' = (\rho'', \theta)$ and $\boldsymbol{\kappa} = (K, \alpha)$ are used. The inner θ integral can be solved using the following Bessel identity:

$$J_o(\kappa\rho) = \frac{1}{2\pi}\int_0^{2\pi} \exp(i\kappa\rho\cos(\theta))d\theta \tag{7.94}$$

where $J_o(\kappa\rho)$ is the Bessel function of the first kind with order zero. This gives us the expression:

$$I = 2\pi\exp(i\boldsymbol{\kappa}\cdot\boldsymbol{\rho})\int_0^\infty \rho'' d\rho'' J_0(\kappa\rho'')\cos\left[\frac{k(\rho'')^2}{2(z - z')}\right] \tag{7.95}$$

[2] Similar steps can be carried out to obtain the expression for phase fluctuations $S(\boldsymbol{r})$.

This integral is equal to the Hankel transform of the cosine term and so the equation (7.95) may be written as:

$$I = \frac{2\pi(z - z')}{k} \exp(i\boldsymbol{\kappa} \cdot \boldsymbol{\rho}) \sin\left[\frac{\kappa^2(z - z')}{2k}\right] \tag{7.96}$$

Therefore, on substituting I in equation (7.91), one can write amplitude (and similarly phase) fluctuations as:

$$\begin{bmatrix} \chi(\boldsymbol{r}) \\ S(\boldsymbol{r}) \end{bmatrix} = \int \exp(i\boldsymbol{\kappa} \cdot \boldsymbol{\rho}) \left\{ k \int_0^z dz' d\nu(\boldsymbol{\kappa}, z') \begin{bmatrix} \sin \\ \cos \end{bmatrix} \left[\frac{\kappa^2(z - z')}{2k}\right] \right\} \tag{7.97}$$

On comparing equations (7.90) and (7.97), a similarity between the two expressions can be observed. The term in the curly brackets in equation (7.97) can be interpreted as the random spectral amplitudes of the amplitude and phase fluctuations and it plays the same role as that played by the random function $\nu(\boldsymbol{\kappa}, z')$ in equation (7.90).

7.3.2.3 Covariance function for the amplitude and phase fluctuations

We know from previous discussions that $\langle \chi(\boldsymbol{r}) \rangle = \langle S(\boldsymbol{r}) \rangle = 0$. It is commonly assumed that the amplitude and phase fluctuations are statistically homogeneous (stationary) in the given z plane. Therefore, one can evaluate the two-dimensional covariance function for the amplitude and phase fluctuations, $B_\chi(\boldsymbol{\rho})$ and $B_S(\boldsymbol{\rho})$, respectively:

$$B_\chi(\boldsymbol{\rho}) = \langle \chi(\boldsymbol{\rho}_1 + \boldsymbol{\rho}, z)\chi^*(\boldsymbol{\rho}_1, z) \rangle \tag{7.98}$$

$$B_S(\boldsymbol{\rho}) = \langle S(\boldsymbol{\rho}_1 + \boldsymbol{\rho}, z)S^*(\boldsymbol{\rho}_1, z) \rangle \tag{7.99}$$

Substituting equation (7.97) into equation (7.98), the expression for $B_\chi(\boldsymbol{\rho})$[3] is obtained as:

$$B_\chi(\boldsymbol{\rho}) = k^2 \int \int e^{i\boldsymbol{\kappa}\cdot(\boldsymbol{\rho}_1 + \boldsymbol{\rho}) - i\boldsymbol{\kappa}'\cdot\boldsymbol{\rho}_1} \int_0^z dz' \int_0^z dz'' \sin\left(\frac{\kappa^2(z - z')}{2k}\right) \sin\left(\frac{\kappa'^2(z - z'')}{2k}\right) \cdot \langle d\nu(\boldsymbol{\kappa}, z') d\nu(\boldsymbol{\kappa}', z'') \rangle \tag{7.100}$$

The ensemble averages of any statistically homogeneous random process are independent of the location at which they are being computed. That is, if we translate our sensors to any other position in the field, and perform the same operation, then the same average properties of the random process should be expected. Therefore, the random amplitude term in the angular brackets should satisfy the relation:

$$\langle d\nu(\boldsymbol{\kappa}, z') d\nu(\boldsymbol{\kappa}', z'') \rangle = \delta(\boldsymbol{\kappa} - \boldsymbol{\kappa}') F_n(\boldsymbol{\kappa}', z' - z'') d^2\boldsymbol{\kappa} d^2\boldsymbol{\kappa}' \tag{7.101}$$

The function $F_n(\boldsymbol{\kappa}', z' - z'')$ is the two-dimensional spectral density of the refractive index fluctuations defined by:

[3] Similar calculations can be performed for $B_S(\boldsymbol{\rho})$.

$$F_n(\boldsymbol{\kappa}', z' - z'') = \int_{-\infty}^{+\infty} d\kappa_z \;\; \Phi_n(\boldsymbol{\kappa}', \kappa_z)\cos[\kappa_z(z' - z'')], \tag{7.102}$$

where $\Phi(\kappa)$ is the three-dimensional spectrum for refractive index fluctuations. Inserting equation (7.101) into equation (7.100) and performing the $\boldsymbol{\kappa}'$ integration, the covariance function $B_\chi(\boldsymbol{\rho})$ has the form:

$$B_\chi(\boldsymbol{\rho}) = \int d^2\boldsymbol{\kappa} e^{i\boldsymbol{\kappa}\cdot\boldsymbol{\rho}}\left\{k^2\int_0^z dz'\int_0^z dz'' \sin\left(\frac{\kappa^2(z - z')}{2k}\right)\sin\left(\frac{\kappa^2(z - z'')}{2k}\right)F_n(\boldsymbol{\kappa}, z' - z'')\right\}. \tag{7.103}$$

Similarly, the covariance function $B_S(\boldsymbol{\rho})$ is given by:

$$B_S(\boldsymbol{\rho}) = \int d^2\boldsymbol{\kappa} e^{i\boldsymbol{\kappa}\cdot\boldsymbol{\rho}}\left\{k^2\int_0^z dz'\int_0^z dz'' \cos\left(\frac{\kappa^2(z - z')}{2k}\right)\cos\left(\frac{\kappa^2(z - z'')}{2k}\right)F_n(\boldsymbol{\kappa}, z' - z'')\right\} \tag{7.104}$$

Therefore, the covariance functions $B_\chi(\boldsymbol{\rho})$ and $B_S(\boldsymbol{\rho})$ are simply the Fourier transform of the quantities inside the curly brackets in equation (7.103) and equation (7.104). These quantities represent the two-dimensional spectral densities of the amplitude and phase fluctuations $F_\chi(\boldsymbol{\kappa}, 0)$ and $F_S(\boldsymbol{\kappa}, 0)$, respectively:

$$F_\chi(\boldsymbol{\kappa}, 0) = k^2\int_0^z dz'\int_0^z dz'' \sin\left(\frac{\kappa^2(z - z')}{2k}\right)\sin\left(\frac{\kappa^2(z - z'')}{2k}\right)F_n(\boldsymbol{\kappa}, z' - z'') \tag{7.105}$$

$$F_S(\boldsymbol{\kappa}, 0) = k^2\int_0^z dz'\int_0^z dz'' \cos\left(\frac{\kappa^2(z - z')}{2k}\right)\cos\left(\frac{\kappa^2(z - z'')}{2k}\right)F_n(\boldsymbol{\kappa}, z' - z'') \tag{7.106}$$

Let us first try to reduce $F_\chi(\boldsymbol{\kappa}, 0)$ and $F_S(\boldsymbol{\kappa}, 0)$ to a simpler form. This can be achieved by making use of the properties of the $F_n(\boldsymbol{\kappa}, z' - z'')$ as first shown by Tatarskii. It is clear from equation (7.102) that $F_n(\boldsymbol{\kappa}, z' - z'')$ is an even function of $z' - z''$, that is

$$F_n(\boldsymbol{\kappa}, z' - z'') = F_n(\boldsymbol{\kappa}, z'' - z') \tag{7.107}$$

which represents isotropic turbulence and so in equations (7.105 and 7.106) vector notation for F_n can be dropped. Next, we make a transformation of variables from z' and z'' to the difference variable $\xi = z' - z''$ and center of mass variable $2\eta = z' + z''$, respectively. The region of integration for the new variables (ξ, η) is a rhombus whose limits are given by the straight lines: $\eta = 0.5\xi$, $\eta = -0.5\xi$, $\eta = z + 0.5\xi$ and $\eta = z - 0.5\xi$, as shown in figure 7.6. Assuming the total propagation length $z = L$ and using the trigonometric identity for the product of two sine functions, the equation for $F_\chi(\kappa, 0)$ reduces to:

$$F_\chi(\kappa, 0) = k^2\int_0^L d\xi \;\; F_n(\kappa, \xi)\int_{\frac{\xi}{2}}^{L-\frac{\xi}{2}} d\eta\left[\cos\left(\frac{\kappa^2\xi}{2k}\right) - \cos\left(\frac{\kappa^2(L - \eta)}{k}\right)\right] \tag{7.108}$$

As $F_n(\kappa, \xi)$ is a function of ξ only, the η integral can be evaluated independently. This gives us:

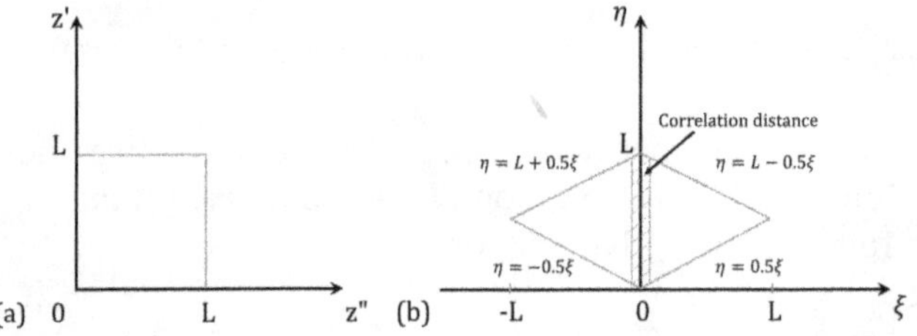

Figure 7.6. Range of integration shown for (a) $z' - z''$ coordinates and (b) $\eta - \xi$ coordinates. Note that the coordinate transformation from $z' - z''$ to $\eta - \xi$ changes the shape of the region of integration from a square to a rhombus. The equations for sides of this rhombus are also given in (b). The shaded region shows the distance over which the correlation is significant.

$$F_\chi(\kappa, 0) = \int_0^L d\xi \ F_n(\kappa, \xi)\left[k^2(L - \xi)\cos\left(\frac{\kappa^2\xi}{2k}\right) + \frac{k^3}{\kappa^2}\sin\left(\frac{\kappa^2(\xi)}{2k}\right) - \frac{k^3}{\kappa^2}\sin\left(\frac{\kappa^2(2L - \xi)}{2k}\right)\right] \tag{7.109}$$

The two-dimensional spectral density $F_n(\kappa, \xi)$ represents the correlation of the refractive index fluctuations n_1 on any two adjacent planes z' and z''. If the separation between these two places is large, then smaller eddies would not intersect both planes and so will contribute negligibly to the correlation. Therefore, only contribution from those turbulent eddies is significant whose spatial size, $l = 2\pi\kappa^{-1}$ is less than or equal to the separation ξ. The implies that the function $F_n(\kappa, z' - z'')$ falls quickly to zero for $\kappa\xi \geqslant 1$. The maximum value of κ ($\kappa_{\max}$) is given by the smallest eddy size or the lower scale of the turbulence l_o. Therefore, the important range of integration is given by $\xi \leqslant \kappa^{-1} \leqslant \kappa_{\max}^{-1}$. We had earlier assumed that the wavelength of the beam is much less than l_o, implying that the quantity $\kappa^2\xi/k \leqslant \lambda/l_o \ll 1$. Apart from these conditions, we also have that the integral above has major contributions only for values of $\xi \ll L$. Since we are interested in correlation of χ and S over such small distances, we can approximate the trigonometric quantities in equation (7.109) as $\cos(\kappa^2\xi/2k) \approx 1$, $\sin(\kappa^2\xi/2k) \approx \kappa^2\xi/2k$ and $\sin(\kappa^2(2L - \xi)/2k) \approx \sin(\kappa^2 L/k)$, and write:

$$F_\chi(\kappa, 0) = \left[k^2L - \frac{k^3}{\kappa^2}\sin\left(\frac{\kappa^2L}{k}\right)\right]\int_0^L F_n(\kappa, \xi) \ d\xi. \tag{7.110}$$

The function $F_n(\kappa, \xi)$ has an appreciable value only over the correlation distance (see figure 7.6), that is, when $\xi \ll L$. So the integration limits in equation (7.110) can be changed to $+\infty$ without much change in the value of the ξ integral. Thus, the final form of the two-dimensional spectral density function for the amplitude fluctuations may be written as:

$$F_\chi(\kappa, 0) = \pi k^2 L\left[1 - \frac{k}{\kappa^2 L}\sin\left(\frac{\kappa^2 L}{k}\right)\right]\Phi_n(\kappa) \tag{7.111}$$

Similarly, the two-dimensional spectral density function for the phase fluctuations is given by:

$$F_S(\kappa, 0) = \pi k^2 L\left[1 + \frac{k}{\kappa^2 L}\sin\left(\frac{\kappa^2 L}{k}\right)\right]\Phi_n(\kappa) \tag{7.112}$$

These expressions for the two-dimensional spectral density of the amplitude and phase fluctuations can be substituted in equations (7.103 and 7.104) to evaluate the covariance functions $B_\chi(\rho)$ and $B_S(\rho)$, respectively. However, before that it is important to understand the various scales associated with $F_\chi(\kappa, 0)$ and $F_S(\rho)$. Consider the two-dimensional spectral density for the amplitude fluctuations $F_\chi(\kappa, 0)$ which is a product of two functions:

(a) The three-dimensional spectrum of the refractive index fluctuations $\Phi_n(\kappa)$, and

(b) The function $f(\kappa) = \left[1 - \frac{k}{\kappa^2 L}\sin\left(\frac{\kappa^2 L}{k}\right)\right]$.

The function $\Phi_n(\kappa)$ is defined for the spatial frequencies lying in the range bounded by the inner and outer scales of turbulence: $\frac{2\pi}{L_0} < \kappa < \frac{2\pi}{l_0}$. For spatial scales which are much smaller than the inner scale l_o (or $\kappa \gg \frac{2\pi}{l_0}$), the value of $\Phi_n(\kappa)$ is negligibly small. Similarly, the function $\Phi_n(\kappa)$ loses its meaning for spatial scales larger than the outer scale of turbulence L_o as for such large eddies the fundamental assumption of an isotropic and homogeneous turbulence is no longer true. Similarly, consider function $f(\kappa)$ for $\frac{\kappa^2 L}{k} \ll 1$ where the sine function can be expanded into a Taylor series:

$$f(\kappa) = 1 - \frac{k}{\kappa^2 L}\left[\left(\frac{\kappa^2 L}{k}\right) - \frac{1}{3!}\left(\frac{\kappa^2 L}{k}\right)^3 + \cdots\right] \approx \frac{1}{6}\frac{\kappa^4 L^2}{k^2}. \tag{7.113}$$

On the other hand, for large values of κ, the sine function oscillates fast, in turn contributing negligibly to $f(\kappa)$ and thus the function $f(\kappa) \to 1$. A characteristic scale $\kappa_1 = \frac{2\pi}{\sqrt{\lambda L}}$ can be defined at which $\sin\left(\frac{\kappa_0^2 L}{k}\right) = 0$ and $f(\kappa) = 1$. Therefore, we can write the function $f(\kappa)$ as:

$$f(\kappa) = \begin{cases} \frac{1}{6}\frac{\kappa^4 L^2}{k^2}, & \kappa < \kappa_1 \\ 1, & \kappa > \kappa_1 \end{cases} \tag{7.114}$$

Depending on the value of the Fresnel length $\sqrt{\lambda L}$ with respect to the l_o and L_o, there are three different cases for the value of two-dimensional spectral density $F_\chi(\kappa, 0)$ can be studied:

(a) $l_o > \sqrt{\lambda L}$

Figure 7.7(a) shows this particular case. It can be seen that the quantity κ_1 is greater than the maximum allowed spatial frequency $\kappa_{\max} \sim \frac{2\pi}{l_0}$ for the

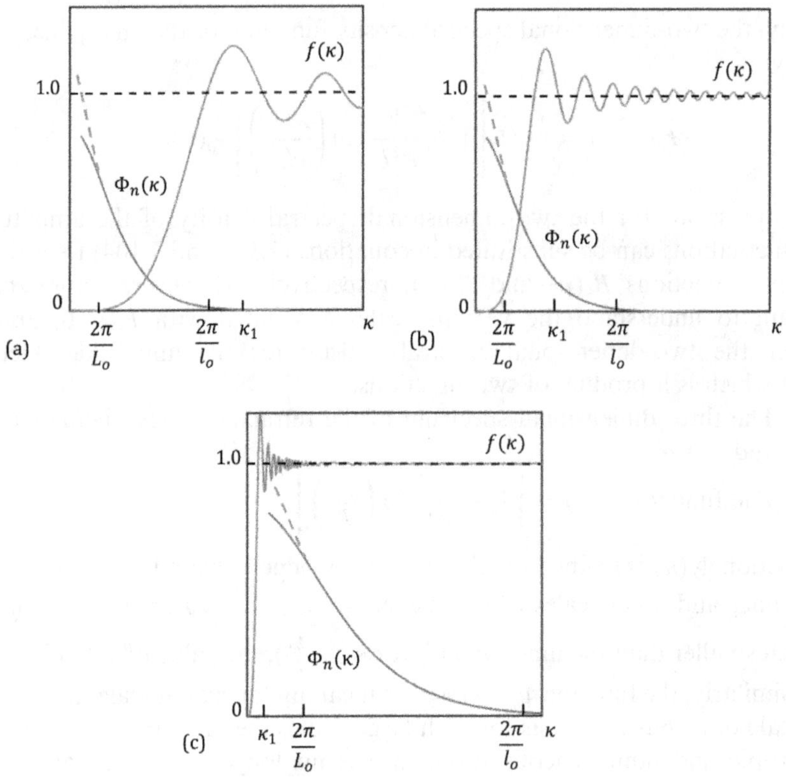

Figure 7.7. The variation of functions $\Phi_n(\kappa)$ and $f(\kappa)$ as a function of the spatial frequency κ are shown for different cases of κ_1, l_o and L_o. The three different cases (a) $l_o > \sqrt{\lambda L}$ (b) $L_o > \sqrt{\lambda L} > l_o$ and (c) $L_o < \sqrt{\lambda L}$ are shown here.

refractive index spectrum $\Phi_n(\kappa)$. The spectrum $\Phi_n(\kappa) \to 0$ in the region where $\kappa > \kappa_{\max}$. From the properties of the function $f(\kappa)$, it can be approximated by $\frac{1}{6}\frac{\kappa^4 L^2}{\kappa^2}$ for all values of $\kappa < \kappa_1$. Thus, we can write the two-dimensional spectral density of the amplitude fluctuations $F_\chi(\kappa, 0)$ as:

$$F_\chi(\kappa, 0) = \begin{cases} \frac{1}{6}\pi L^3 \kappa^4 \Phi_n(\kappa), & \kappa < \kappa_{\max} \\ 0, & \kappa > \kappa_{\max} \end{cases} \tag{7.115}$$

The product of functions $\Phi_n(\kappa)$ and $f(\kappa)$ will have a maximum value near $\kappa_{\max}$. This means that for the case when $l_o > \sqrt{\lambda L}$, the refractive index inhomogeneities with sizes of the order of l_o will have the greatest effect on the amplitude fluctuations. The covariance function $B_\chi(\rho)$ in this case would therefore have a correlation length of the order of l_o.

(b) $L_o > \sqrt{\lambda L} > l_o$

In this case, the product of the functions $\Phi_n(\kappa)$ and $f(\kappa)$ has a maximum near the point κ_1 and becomes zero for $\kappa > \kappa_{\max}$ as clearly seen in figure 7.7 (b). Therefore, $F_\chi(\kappa, 0)$ can be written as:

$$F_\chi(\kappa, 0) = \begin{cases} \frac{1}{6}\pi L^3 \kappa^4 \Phi_n(\kappa), & \kappa < \kappa_l \\ \pi k^2 L \Phi_n(\kappa), & \kappa_l < \kappa < \kappa_{\max} \\ 0, & \kappa > \kappa_{\max} \end{cases} \tag{7.116}$$

In this case, refractive index inhomogeneities of the spatial size $\sqrt{\lambda L}$ cause the major amplitude fluctuations in the light beam. At the final detector plane situated at $x = L$, the correlation function $B_\chi(\boldsymbol{\rho})$ would show a correlation distance of the order of $\sqrt{\lambda L}$ or the Fresnel length for the particular propagation geometry. Notice that in figure 7.7(a)–(b), the function $f(\kappa)$ is near zero for all frequencies $\kappa < \kappa_{\min}$ where $\kappa_{\min} \sim \frac{2\pi}{L_o}$.

(c) $L_o < \sqrt{\lambda L}$

In this case, the major influence on the amplitude fluctuations comes from refractive index inhomogeneities with spatial sizes between L_o and $\sqrt{\lambda L}$. However, the refractive index eddies whose sizes are greater than the outer scale L_o cannot be considered isotropic or homogeneous. Therefore, the correlation function between the refractive index inhomogeneities is now no longer independent of the location of the observation points. Therefore, now the spectral density $\Phi_n(\kappa)$ is no longer defined in this region. The $\Phi_n(\kappa)$ has meaning only in the region when $\kappa \geqslant \kappa_{\min}$. So we can write the function $F_\chi(\kappa, 0)$ as:

$$F_\chi(\kappa, 0) = \pi k^2 L \Phi_n(\kappa), \quad \kappa > \kappa_{\min} \tag{7.117}$$

In the case of $F_S(\kappa, 0)$, the function $f(\kappa) = \left[1 + \frac{k}{\kappa^2 L}\sin\left(\frac{\kappa^2 L}{k}\right)\right]$ has a maximum at $\kappa = 0$. This means that the refractive index eddies of scale sizes L_o greatly influence the phase fluctuations, which means that in order to get stable statistics, long measurement times are required. However, these eddies are nonstationary in space as the atmosphere is dynamically changing at millisecond timescales, thus corrupting the measured data. Thus, it is not practical to measure the covariance function $B_S(\boldsymbol{\rho})$. Instead, the phase structure function $D_S(\rho)$ defined as:

$$D_S(\rho) = 2[B_S(0) - B_S(\rho)] \tag{7.118}$$

offers a more physically useful measure of the phase fluctuations. The covariance functions $B_\chi(\rho)$ and $B_S(\rho)$ can be obtained by taking the Fourier transform of its respective spectral density functions $F_\chi(\kappa, 0)$ and $F_S(\kappa, 0)$. We can substitute $F_\chi(\kappa, 0)$ and $F_S(\kappa, 0)$ from equations (7.111–7.112) into equations (7.103–7.104) to obtain:

$$B_{\chi, S}(\rho) = \pi k^2 L \int d^2\boldsymbol{\kappa} e^{i\boldsymbol{\kappa}\cdot\boldsymbol{\rho}}\left[1 \mp \frac{k}{\kappa^2 L}\sin\left(\frac{\kappa^2 L}{k}\right)\right]\Phi_n(\kappa). \tag{7.119}$$

We can represent $\boldsymbol{\kappa}$ in polar coordinates and substitute $\boldsymbol{\kappa} = (\kappa, \theta)$ and $d^2\boldsymbol{\kappa} = \kappa dk d\theta$ in the above equation. Further, as we are dealing with isotropic and homogeneous turbulence, we use the Bessel identity (see equation (7.94)) to evaluate the θ integral and get:

$$B_{\chi,S}(\rho) = 2\pi^2k^2L\int_0^\infty \kappa J_0(\kappa\rho)\left[1 \mp \frac{k}{\kappa^2L}\sin\left(\frac{\kappa^2L}{k}\right)\right]\Phi_n(\kappa)d\kappa. \tag{7.120}$$

The phase structure function $D_s(\rho)$ can thus be given as:

$$D_S(\rho) = 4\pi^2k^2L\int_0^\infty \kappa\left[1 + \frac{k}{\kappa^2L}\sin\left(\frac{\kappa^2L}{k}\right)\right][1 - J_0(\kappa\rho)]\Phi_n(\kappa)d\kappa. \tag{7.121}$$

Similar calculations for the log-amplitude structure function $D_\chi(\rho)$ yield the relation,

$$D_\chi(\rho) = 4\pi^2k^2L\int_0^\infty \kappa\left[1 - \frac{k}{\kappa^2L}\sin\left(\frac{\kappa^2L}{k}\right)\right][1 - J_0(\kappa\rho)]\Phi_n(\kappa)d\kappa. \tag{7.122}$$

The total wave structure function is equal to the sum of the log-amplitude and phase structure functions,

$$D(\rho) = D_\chi(\rho) + D_S(\rho), \tag{7.123}$$

which for the case of unbounded plane waves is obtained from equations (7.121) and (7.122) as,

$$D(\rho) = 8\pi^2k^2L\int_0^\infty \kappa\ \Phi_n(\kappa)[1 - J_0(\kappa\rho)]d\kappa. \tag{7.124}$$

The particular form for the amplitude covariance function and the phase structure function is obtained by plugging the desired refractive index spectrum into the above equations. Substituting the Kolmogorov spectrum for $\Phi_n(\kappa)$, we get:

$$B_\chi(\rho) = 2\pi^2(0.033)C_n^2k^2L\int_0^\infty d\kappa\ J_0(\kappa\rho)\left[1 - \frac{k}{\kappa^2L}\sin\left(\frac{\kappa^2L}{k}\right)\right]\kappa^{-8/3}, \tag{7.125}$$

$$D_S(\rho) = 4\pi^2(0.033)C_n^2k^2L\int_0^\infty d\kappa\left[1 + \frac{k}{\kappa^2L}\sin\left(\frac{\kappa^2L}{k}\right)\right][1 - J_0(\kappa\rho)]\kappa^{-8/3}. \tag{7.126}$$

Assuming that $l_o \gg \sqrt{\lambda L}$, we can calculate the variance of the amplitude fluctuations σ_χ^2 from the covariance function as:

$$\sigma_\chi^2 = B_\chi(\rho = 0) = 0.31C_n^2k^{7/6}L^{11/6}. \tag{7.127}$$

The power law for phase structure function for the different conditions is then obtained as:

$$D_s(\rho) = \begin{cases} 1.46C_n^2 k^2 L\rho^{5/3}, & \text{when } l_o \ll \rho \ll \sqrt{\lambda L}, \\ 2.91C_n^2 k^2 L\rho^{5/3}, & \text{when } \sqrt{\lambda L} \ll \rho \ll L_o. \end{cases} \tag{7.128}$$

7.3.2.4 Mutual coherence function

The mutual coherence function or the second moment for the field at some propagation distance L is given by the ensemble average,

$$\Gamma_2(\boldsymbol{r}_1, \boldsymbol{r}_2, L) = \langle U(\boldsymbol{r}_1, L) \;\; U^*(\boldsymbol{r}_2, L)\rangle, \tag{7.129}$$

where $U^*(\boldsymbol{r}, L)$ denotes the complex conjugate field. We substitute the field $U(\boldsymbol{r}, L)$ from equation (7.51) to get,

$$\Gamma_2(\boldsymbol{r}_1, \boldsymbol{r}_2, L) = U_0(\boldsymbol{r}_1, L) \;\; U_0^*(\boldsymbol{r}_2, L) \;\; \langle \exp[\psi(\boldsymbol{r}_1) + \psi^*(\boldsymbol{r}_2)]\rangle. \tag{7.130}$$

The mutual coherence function, when evaluated at identical observation points $\boldsymbol{r}_1 = \boldsymbol{r}_2 = \boldsymbol{r}$ gives the mean irradiance value,

$$\langle I(\boldsymbol{r}, L)\rangle = \Gamma_2(\boldsymbol{r}, \boldsymbol{r}, L). \tag{7.131}$$

The mean irradiance can be used to predict the atmosphere-induced beam spreading [3]. For the case of an infinite plane wave with no wavefront curvature, the mutual coherence function is given by the expression,

$$\Gamma_2(\rho, L) = \exp\left[-4\pi^2 k^2 L \int_0^\infty \kappa \;\; \Phi_n(\kappa)[1 - J_o(\kappa\rho)]d\kappa\right], \tag{7.132}$$

where $\rho = |\boldsymbol{r}_1 - \boldsymbol{r}_2|$ is the separation distance between two points on the wavefront. The mutual coherence function can also be used to predict the spatial coherence radius at the detector plane. The modulus of the complex degree of coherence (DOC), which provides information regarding the loss of spatial coherence of an initially coherent beam is given by,

$$\mathrm{DOC}(\boldsymbol{r}_1, \boldsymbol{r}_2, L) = \frac{|\Gamma_2(\boldsymbol{r}_1, \boldsymbol{r}_2, L)|}{\sqrt{\Gamma_2(\boldsymbol{r}_1, \boldsymbol{r}_1, L) \;\; \Gamma_2(\boldsymbol{r}_1, \boldsymbol{r}_1, L)}} \tag{7.133}$$

$$= \exp\left[-\frac{1}{2}D(\boldsymbol{r}_1, \boldsymbol{r}_2, L)\right] \tag{7.134}$$

Here, $D(\boldsymbol{r}_1, \boldsymbol{r}_2, L)$ is the wave structure function given by equation (7.124). The spatial coherence radius ρ_o (measured in meters) is defined by the separation distance at which the DOC reduces to value $1/e$ i.e. when $D(\rho, L) = 2$. For the unbounded plane wave, we get

$$\rho_o = \begin{cases} \left[1.46k^2 \int_0^L C_n^2(z)dz\right]^{-3/5}, & \text{when } l_o \ll \rho_o \ll L_o, \\ \left[1.46C_n^2 k^2 L\right]^{5/3}, & \text{when } l_o \ll \rho_o \ll L_o \text{ and } C_n^2 = \text{Constant}. \end{cases} \tag{7.135}$$

Sometimes, the spatial coherence is defined through atmospheric coherence width (r_o) which is equal to $r_o = 2.1\rho_o$. This term, also commonly known as the Fried parameter, was first introduced by Fried [26, 27] who showed that it is an important measure of the performance of an imaging system. In the presence of turbulence, the angular extent (seeing angle) of the far-field diffraction pattern is given as λ/r_o. The Fried parameter can be interpreted as the aperture size beyond which further increases in diameter result in no further increase in resolution. The Fried parameter for plane waves is given as,

$$r_{op} = \left[0.42k^2 \int_0^L C_n^2(z)dz\right]^{-3/5}. \tag{7.136}$$

7.3.2.5 Cross-coherence function

The fluctuations in the irradiance of the field are described by the cross-coherence function or the fourth-order moment of the field,

$$\Gamma_4(\boldsymbol{r}_1, \boldsymbol{r}_2, \boldsymbol{r}_3, \boldsymbol{r}_4, L) = \langle U(\boldsymbol{r}_1, L)U^*(\boldsymbol{r}_2, L)U(\boldsymbol{r}_3, L)U^*(\boldsymbol{r}_4, L)\rangle \tag{7.137}$$

which can be expressed in the form,

$$\begin{aligned}\Gamma_4(\boldsymbol{r}_1, \boldsymbol{r}_2, \boldsymbol{r}_3, \boldsymbol{r}_4, L) = \; & U_0(\boldsymbol{r}_1, L)U_0^*(\boldsymbol{r}_2, L)U_0(\boldsymbol{r}_3, L)U_0^*(\boldsymbol{r}_4, L) \times \\ & \langle \exp[\psi(\boldsymbol{r}_1) + \psi^*(\boldsymbol{r}_2) + \psi(\boldsymbol{r}_3) + \psi^*(\boldsymbol{r}_4)]\rangle\end{aligned} \tag{7.138}$$

By setting $\boldsymbol{r}_1 = \boldsymbol{r}_2 = \boldsymbol{r}_3 = \boldsymbol{r}_4 = \boldsymbol{r}$, the fourth-order coherence function yields the second moment of irradiance,

$$\langle I^2(\boldsymbol{r}, L)\rangle = \Gamma_4(\boldsymbol{r}, \boldsymbol{r}, \boldsymbol{r}, \boldsymbol{r}, L) \tag{7.139}$$

The covariance function of irradiance is defined by the normalized quantity,

$$B_I(\boldsymbol{r}_1, \boldsymbol{r}_2, L) = \frac{\Gamma_4(\boldsymbol{r}_1, \boldsymbol{r}_1, \boldsymbol{r}_2, \boldsymbol{r}_2, L) - \Gamma_2(\boldsymbol{r}_1, \boldsymbol{r}_1, L)\Gamma_2(\boldsymbol{r}_2, \boldsymbol{r}_2, L)}{\Gamma_2(\boldsymbol{r}_1, \boldsymbol{r}_1, L)\Gamma_2(\boldsymbol{r}_2, \boldsymbol{r}_2, L)} \tag{7.140}$$

$$= \frac{\Gamma_4(\boldsymbol{r}_1, \boldsymbol{r}_1, \boldsymbol{r}_2, \boldsymbol{r}_2, L)}{\Gamma_2(\boldsymbol{r}_1, \boldsymbol{r}_1, L)\Gamma_2(\boldsymbol{r}_2, \boldsymbol{r}_2, L)} - 1, \tag{7.141}$$

which for the case $\boldsymbol{r}_1 = \boldsymbol{r}_2 = \boldsymbol{r}$ gives us the scintillation index,

$$\sigma_I^2(\boldsymbol{r}, L) = \frac{\langle I^2(\boldsymbol{r}, L)\rangle}{\langle I(\boldsymbol{r}, L)\rangle} - 1. \tag{7.142}$$

Scintillation refers to the temporal and spatial fluctuations in the irradiance of a beam on propagation through atmospheric turbulence. When the log-amplitude variance σ_χ^2 is sufficiently small, i.e., $\sigma_\chi^2 \ll 1$, then

$$\sigma_I^2(\boldsymbol{r}, L) = \exp\left[4\sigma_\chi^2(\boldsymbol{r}, L)\right] - 1 \tag{7.143}$$

Table 7.1. Classification of refractive index fluctuations on the basis of Rytov variance.

Fluctuation condition	Rytov variance value
Weak	$\sigma_I^2 < 1$
Moderate	$\sigma_I^2 \sim 1$
Strong	$\sigma_I^2 > 1$
Saturation	$\sigma_I^2 \to \infty$

$$\cong\ 4\sigma_\chi^2(\boldsymbol{r},\ L). \tag{7.144}$$

The scintillation index can be expressed as a sum of radial and longitudinal components,

$$\sigma_I^2(\boldsymbol{r},\ L) = \sigma_{I,\ \mathrm{rad}}^2(\boldsymbol{r},\ L) + \sigma_{I,\ \mathrm{long}}^2(\boldsymbol{r},\ L). \tag{7.145}$$

The radial component of the scintillation index, $\sigma_{I,\ \mathrm{rad}}^2(\boldsymbol{r},\ L)$ is zero at the beam center $r = 0$. This component is also zero for the case of an infinite plane or spherical wave. The longitudinal component of the scintillation index $\sigma_{I,\ \mathrm{long}}^2(\boldsymbol{r},\ L)$ is constant across the beam cross-section in any z plane. The longitudinal component corresponds to the on-axis scintillation index and is also written as $\sigma_I^2(0,\ L)$. Using equation (7.127) as the form of σ_χ^2, the scintillation index for a plane wave is given as,

$$\sigma_R^2 = 1.23 C_n^2 k^{7/6} L^{11/6}. \tag{7.146}$$

This quantity is also commonly known as Rytov variance [3]. The Rytov variance value can be used as a theoretical measure to distinguish between weak or strong refractive fluctuation regimes, as shown in table 7.1.

7.3.2.6 Validity of the Rytov method

The assumption that $|\nabla\psi_1| \ll |\nabla\psi_0|$ leads to important restrictions on the range of validity of the Rytov method. When the Rytov method was first developed, it appeared to give results which were in good agreement with the experimental data, that were measured over atmospheric propagation paths of less than 1 km. This led to the belief that this method gave amplitude and phase results for a much greater range of validity in terms of C_n^2 and propagation path compared to the Born approximation. Later, studies by several researchers [28–31] challenged this notion of the extended range of validity of this method. It is now well established that the results obtained through the Rytov approach are valid only when $\sigma_R^2 \leqslant 0.3$ [1, 32]. Therefore, it can be applied to very short distances when the turbulence is strong. However, at night, when the value of C_n^2 may drop up to two orders of magnitude, this method may sometimes provide good estimates even for propagation paths of 100 km. The method would also be applicable for longer ranges when the light wavelength is an order of magnitude or more than the optical range. Also, some experimental work suggests that the in strong integrated turbulence, the phase statistics can have a larger range of validity than log-amplitude statistics [18, 32, 33]. Newer methods like parabolic equation method, the diagram method, coherence theory-based approach and the Markov approximation have also been explored in this regard [5]. Recently, the extended Huygens–Fresnel (eHF) method [34–36] has also emerged as a popular approach for studying propagation of a beam through random media like atmospheric turbulence. We will describe the eHF approach in some detail in the next section.

7.4 Extended Huygens–Fresnel integral approach

According to the Huygens construction, in the propagation of any beam through a vacuum, every point on the wavefront can be considered as a center of a secondary disturbance which gives rise to secondary wavelets. At any later instant, the wavefront may be regarded as the envelope of these wavelets. Fresnel supplemented Huygens' principle by providing an explanation for the process of diffraction by stating that these secondary wavelets mutually interfere. The combination of Huygens' construction with the principle of interference is together called the *Huygens–Fresnel principle*. It was shown independently by Feizulin and Kravtsov [35] in 1967 and by Lutomirski and Yura [34] in 1971 that one can develop an extension of the Huygens–Fresnel principle for propagation in a weakly inhomogeneous medium. They showed that for a refractive medium, the secondary wavefront will again be determined by the envelope of spherical wavelets originating from the primary wavefront. However, now each wavelet will be determined by its propagation through the inhomogeneous medium. Starting from the stochastic Helmholtz equation, Lutomirski and Yura followed a method very similar to the integral theorem of Helmholtz and Kirchhoff used in free-space diffraction theory and showed that the field at any arbitrary point P_0 inside the medium can be expressed in terms of the boundary values of the wave on any closed surface surrounding that point. Let the turbulent medium occupy a volume V which is bounded by a surface S and the refractive index variation inside the volume is given by equation (7.6). Then, within the volume V, let a point source be situated at P_0 which produces a field $G(\boldsymbol{P}_1, \boldsymbol{P}_0)$ at any point P_1, then using equation (7.8) one obtains:

$$[\nabla^2 + k^2 n^2(\boldsymbol{r})] \ G(\boldsymbol{P}_1, \boldsymbol{P}_0) = -4\pi\delta(|\boldsymbol{P}_1 - \boldsymbol{P}_0|) \tag{7.147}$$

where δ is the Dirac delta function. Multiply equation (7.8) by G and equation (7.147) by U and subtract them to yield,

$$G(\boldsymbol{P}_1, \boldsymbol{P}_0)\nabla^2 U - U\nabla^2 G(\boldsymbol{P}_1, \boldsymbol{P}_0) = 4\pi\delta(|\boldsymbol{r} - \boldsymbol{P}|)U \tag{7.148}$$

If the field U possesses continuous first- and second-order derivatives inside and on S, then one can integrate equation (7.148) over V and apply Green's theorem,

$$U(\boldsymbol{P}_0) = \frac{1}{4\pi} \iint_S (U\nabla G - G\nabla U) \cdot d\boldsymbol{A} \tag{7.149}$$

Here, dA represents a surface element with its normal directed into V. We want to find the complex field distribution at some arbitrary point in a turbulent medium when the optical field specified over a finite surface like the aperture, is available. Therefore, let a monochromatic optical disturbance pass through an aperture A in an opaque screen and propagate into the turbulent medium. The aperture is assumed to be large compared to the wavelength of light but smaller than the distance between the observation point P_0 and the opaque screen. Now again following arguments analogous to the free-space Kirchhoff formulation of diffraction through a planar screen, equation (7.149) is reduced into an integral over an aperture of area Σ,

$$U(\boldsymbol{P}_0) = \frac{1}{4\pi}\iint_{\Sigma}(U\nabla G - G\nabla U)\cdot d\boldsymbol{A} \tag{7.150}$$

Let the optical disturbance propagate along the z-axis and the aperture A lies in a plane normal to this direction. The unit vector normal to the beam wavefront in the aperture is denoted by $\hat{n}$, then $\partial U/\partial z \approx ikU(\hat{n}\cdot\hat{z})$. Next, the function G is chosen to be of the form $G = \exp(ikr_{01} + \psi)/r_{01}$ where r_{01} is the distance between the point P_0 and the elemental area dA in the aperture, given by $s = |\boldsymbol{P}_0 - \boldsymbol{P}_1|$. Following Kirchhoff treatment and using the assumption that $s \gg \lambda$, one can write,

$$U(\boldsymbol{P}_0) = \frac{ik}{4\pi}\iint_{\Sigma} U(\boldsymbol{P}_1)\frac{\exp(ikr_{01} + \psi)}{r_{01}}\left[\hat{r}_{01}\cdot\hat{z} - \hat{n}.\,\hat{z} + \frac{\nabla\psi}{ik}\cdot\hat{z}\right]d^2\boldsymbol{r} \tag{7.151}$$

where $\hat{r}_{01} = (\boldsymbol{P}_0 - \boldsymbol{P}_1)/|\boldsymbol{P}_0 - \boldsymbol{P}_1|$ and $d^2\boldsymbol{r}$ is an element of area at the point P_1 in the aperture. In the geometric optics regime, one can write $\psi \sim ik\int n_1 ds$, then $|(\nabla\psi/ik)\cdot\hat{z}| = |n_1(\hat{s}\cdot\hat{z})|$ where $n_1 \sim 10^{-6}$. The third term in the integral is neglected as it is very small compared to unity for all distances of interest [34]. The area Σ can be replaced with a portion of the incident wavefront which fills the aperture. Over this new region of integration A, the derivative of the field U in the direction normal to the surface is $\partial U/\partial z \leqslant ikU$. This implies that equation (7.150) can be written as,

$$U(\boldsymbol{P}_0) = \int_A K(h)G(\boldsymbol{P}_1', \boldsymbol{P}_0)U(\boldsymbol{r}')d^2\boldsymbol{r}' \tag{7.152}$$

where $K(h) = (-i/2\lambda)(1 + \cos(h))$ and $h = \pi - \cos^{-1}(\hat{n}\cdot\hat{z})$. Remember that $G(\boldsymbol{P}_1, \boldsymbol{P}_0)$ was defined as the field at point P_1 due to a point source situated at P_0. However, it was shown by Yura that the field and source points can be interchanged as $G(\boldsymbol{P}_1, \boldsymbol{P}_0) = G(\boldsymbol{P}_0, \boldsymbol{P}_1)$ [34]. This gives the extended form of Huygens–Fresnel principle for an inhomogeneous medium where the element of area dA of the wavefront contributes to the field at point P_0. Therefore, if one knows how a spherical wave propagates in a given medium, the response to a random disturbance in the aperture can be determined.

The most commonly used form for the extended Huygens–Fresnel principle is obtained by further assuming paraxially propagating waves and using the Fresnel diffraction as the free-space propagator. In that case, the eHF principle is given by:

$$U(\boldsymbol{r}_1) = -\frac{ik}{2\pi L}\exp(ikL)\iint_{-\infty}^{+\infty} U(\boldsymbol{r}_0)\exp\left[\left(\frac{ik\,|\boldsymbol{r}_1 - \boldsymbol{r}_0|^2}{2L}\right) + \psi(\boldsymbol{r}_1, \boldsymbol{r}_0)\right] d^2 r_0 \quad (7.153)$$

However, it was recently shown by Charnotskii [36] that the eHF principle is actually a rigorous consequence of the Helmholtz wave equation and it requires no additional assumptions like (a) the paraxial approximation or (b) weak turbulence. In this case, the kernel of the operator is not the Green's function of the original Helmholtz equation but rather the doubled derivative of the Green's function along the propagation direction.

In equation (7.153), the randomness of the medium is contained in the complex phase term $\psi(\boldsymbol{r}_1, \boldsymbol{r}_0)$ which is being added to each spherical wave. This is a very general description of the turbulent effects on the propagation. The usual practice is to use Rytov approximation to describe the phase term $\psi(\boldsymbol{r}_1, \boldsymbol{r}_0)$,

$$\exp(\psi) = \exp(\psi_1 + \psi_2) \quad (7.154)$$

where ψ_1 and ψ_2 are the first- and second-order Rytov approximation terms. This method is unreliable when the complex phase disturbances $\langle\psi\rangle^2 > 1$. Charnotskii has described the various validity regions of the eHF principle [36]. He has suggested that when the phase term ψ is represented through equation (7.154), then the resultant field violates the energy-conservation principle and in that case the accuracy of the scintillation calculations is uncertain. This form of ψ fetches accurate results for the fourth moment only under weak scintillation conditions.

References

[1] Strohbehn J W 1968 Line-of-sight wave propagation through the turbulent atmosphere *Proc. IEEE* **56** 1301–8 8

[2] Tatarskii V I 1961 *Wave Propagation in a Turbulent Medium* ed R A Silverman (New York: McGraw-Hill)

[3] Andrews L C and Phillips R L 2005 *Laser Beam Propagation through Random Media* (Bellingham, WA: SPIE Press Book)

[4] Wheelon A D 2001 *Electromagnetic Scintillation* **vol 1** (Cambridge: Cambridge University Press)

[5] Strohbehn J W 1978 *Laser Beam Propagation in the Atmosphere* >Topics in Applied Physics vol 25 *(Berlin: Springer)*

[6] Mandel L and Wolf E 1995 *Optical Coherence and Quantum Optics* (Cambridge: Cambridge University Press)

[7] Roggemann M C and Welsh B M 1996 *Imaging Through Turbulence* (Boca Raton, FL: CRC Press)

[8] Goodman J W 2015 *Statistical Optics* 2nd edn (New York: Wiley)

[9] Champagne F H, Friehe C A, LaRue J C and Wynagaard J C 1977 Flux measurements, flux estimation techniques, and fine-scale turbulence measurements in the unstable surface layer over land *J. Atmos. Sci.* **34** 515–30 03

[10] Williams R M and Paulson C A 1977 Microscale temperature and velocity spectra in the atmospheric boundary layer *J. Fluid Mech.* **83** 547–67

[11] Oboukhov A M 1949 Structure of the temperature field in turbulent flow *Izv. Akad. Nauk. SSSR, Ser. Geogr. i Geofiz* **13** 58–69

[12] Corrsin S 1951 On the spectrum of isotropic temperature fluctuations in an isotropic turbulence *J. Appl. Phys.* **22** 469–73

[13] Hill R J 1978 Models of the scalar spectrum for turbulent advection *J. Fluid Mech.* **88** 541–62

[14] Hill R J and Clifford S F 1978 Modified spectrum of atmospheric temperature fluctuations and its application to optical propagation *J. Opt. Soc. Am.* **68** 892–9 7

[15] Churnside J H and Lataitis R J 1990 Wander of an optical beam in the turbulent atmosphere *Appl. Opt.* **29** 926–30 3

[16] Andrews L C 1992 An analytical model for the refractive index power spectrum and its application to optical scintillations in the atmosphere *J. Mod. Opt.* **39** 1849–53

[17] Tatarskii V I 1967 Depolarization of light by turbulent atmospheric inhomogeneities *Radiophys. Quantum Electron.* **10** 987–8

[18] Strohbehn J and Clifford S 1967 Polarization and angle-of-arrival fluctuations for a plane wave propagated through a turbulent medium *IEEE Trans. Antennas Propag.* **15** 416–21

[19] Saleh A A M 1967 An investigation of laser wave depolarization due to atmospheric transmission *IEEE J. Quantum Electron.* **3** 540–3

[20] Cox D C, Arnold H W and Hoffman H H 1981 Observations of cloud-produced amplitude scintillation on 19- and 28-GHz Earth-space paths *Radio Sci.* **16** 885–907

[21] Rytov S M 1937 Diffraction of light by ultrasonic waves *Izv. Akad. Nauk SSSR Ser. Fiz. (Bull. Acad. Sci. USSR Phys. Ser.)* **2** 223–59

[22] Obukhov A M 1953 Effect of weak inhomogeneities in the atmosphere on sound and light propagation *Izv. Akad. Nauk SSSR Ser. Geophys.* **2** 155–65

[23] Tatarskii V I 1971 *The effects of the Turbulent Atmosphere on Wave Propagation* (Washington, D.C., Springfield: National Oceanic and Atmospheric Administration, U.S. Department of Commerce and the National Science Foundation)

[24] Yura H T, Sung C C, Clifford S F and Hill R J 1983 Second-order Rytov approximation *J. Opt. Soc. Am.* **73** 500–2

[25] Yura H T 1969 Optical propagation through a turbulent medium *J. Opt. Soc. Am.* **59** 111–2

[26] Fried D L 1966 Optical resolution through a randomly inhomogeneous medium for very long and very short exposures *J. Opt. Soc. Am.* **56** 1372–9 10

[27] Fried D L 1967 Optical heterodyne detection of an atmospherically distorted signal wave front *Proc. IEEE* **55** 57–77

[28] Brown W P 1967 Validity of the Rytov approximation* *J. Opt. Soc. Am.* **57** 1539–42 12

[29] Lee R W and Harp J C 1969 Weak scattering in random media, with applications to remote probing *Proc. IEEE* **57** 375–406

[30] Tatarski V I 1962 Second approximation in the problem of wave propagation in random medium *Izv. VUZ- Radiofiz.* **5** 490

[31] Hufnagel R E and Stanley N R 1964 Modulation transfer function associated with image transmission through turbulent media *J. Opt. Soc. Am.* **54** 52–61 1

[32] Fante R L 1975 Electromagnetic beam propagation in turbulent media *Proc. IEEE* **63** 1669–92 12

[33] Gurvich A S, Kallistratova M A and Time N S 1968 Fluctuations in the parameters of a light wave from a laser during propagation in the atmosphere *Radiophys. Quantum Electron.* **11** 771–6

[34] Lutomirski R F and Yura H T 1971 Propagation of a finite optical beam in an inhomogeneous medium *Appl. Opt.* **10** 1652–8 7

[35] Feizulin Z I and Kravtsov Y A 1967 Broadening of a laser beam in a turbulent medium Radiophys *Quantum Electron.* **10** 33–5 1

[36] Charnotskii M 2015 Extended Huygen's Fresnel principle and optical waves propagation in turbulence: discussion *J. Opt. Soc. Am.* A **32** 1357–65 7

IOP Publishing

Orbital Angular Momentum States of Light (Second Edition)
Propagation through atmospheric turbulence
Kedar Khare, Priyanka Lochab and Paramasivam Senthilkumaran

Chapter 8

Numerical simulation of laser beam propagation through turbulence

This chapter provides a discussion of numerical methods used in the current literature for simulating realistic turbulence conditions. The approximate theoretical methods presented in the previous chapter provide a good understanding of the underlying physics but have proved to be inadequate in realistic turbulence conditions and propagation distances of at least a few km as required in a number of applications of current interest. This chapter discusses the prevailing methodologies for numerical simulation of beam propagation through atmospheric turbulence and also provides an example code base (in appendix A) that a reader can readily implement in their own work in testing propagation of arbitrarily designed beam profiles.

8.1 Need for specialized numerical methods for beam propagation

In the last chapter, we provided a detailed discussion of the various theoretical methodologies that have been developed for solving the stochastic Helmholtz equation [1]. These analytical methods were based on the approximation that the wavelength of the optical signal is significantly shorter than the typical size of the turbulent eddies in the atmosphere [2]. So the wave undergoes only forward scattering at small angles with respect to the propagation axis. The earlier perturbation series methods like Born's and Rytov's methods [3–8] expressed the statistics of the field $U(\boldsymbol{r})$ in terms of the statistics of the fluctuating part $n_1(\boldsymbol{r})$ of the refractive index. However, these methods are valid as long as the intensity fluctuations in the beam are smaller than $\sim$0.1 times the mean intensity value [9]. This severely limited their usability to small propagation distances and low-to-moderate turbulence strengths. To extend the range of the validity of the wave field solutions into the strong intensity fluctuation regimes, newer approaches like the diagram method, Markov method, and the local method of small approximations [7]

doi:10.1088/978-0-7503-5959-7ch8

were developed. These methods were useful in providing relations for the first and second moments of the field; however, the equations describing higher-order moments remained intractable with limited asymptotic solutions for the fourth moment of the field [6, 10, 11]. Recently, newer asymptotic schemes for determining intensity statistics for waves in turbulence with pure power-law spectra have been proposed [12, 13]. However, a major issue with these asymptotic schemes is that they assume unrealistically strong levels of turbulence. Another popular approach is the use of the extended Huygens–Fresnel principle for understanding the wave propagation phenomena and developing a heuristic theory for propagation through the atmosphere [14–16]. This method has been able to provide many useful insights, but the results obtained are mainly qualitative in nature, and the accuracy of the higher-order moments is not yet ascertained. Therefore, in between the weak turbulence regime where perturbation methods work and unrealistically strong turbulence regime where asymptotic solutions are provided, an entire range of realistic moderate-to-strong turbulence regions exist, which is not entirely understood by the current wave propagation theories. In the absence of satisfactory theoretical description for the intensity fluctuations in this region, numerical simulations offer an attractive alternative approach that has proved to be valuable in practice. Figure 8.1 provides a comparison of the range of validity of various methods for laser propagation in turbulence.

Numerical simulation studies offer various advantages over the analytical methods for solving the stochastic wave equation. When applicable, the theoretical methods usually require lengthy derivations and complicated calculations of moments of the field. On the other hand, numerical simulations can provide an easier and practical way to tackle beams with arbitrary amplitude, phase and polarization profiles. Numerical simulations can also be used to predict higher moments of the field, which can provide some basis for comparison with future analytic studies and experiments. Another drawback of analytical computations is that they only yield statistical

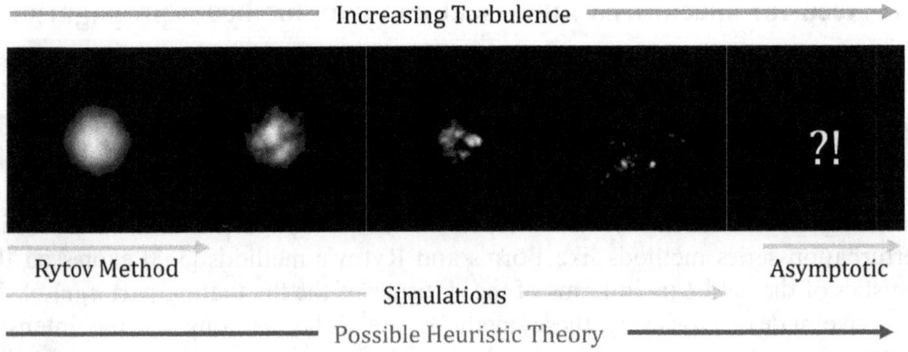

Figure 8.1. The range of validity of the Rytov method, numerical simulation studies, and asymptotic solutions with respect to the increasing turbulence strength are illustrated. In the absence of any heuristic theory, which can describe the experimentally observed results, simulation studies offer a practical solution. Figure layout adapted from [9]. Beam profiles are simulated using the numerical methods described later in this chapter.

quantities like the average irradiance and variance. However, in many applications like adaptive optics [17] and imaging [18], the amplitude and phase profile of the instantaneous optical beam is of interest rather than its statistical behavior. These systems require understanding the evolution of the beam from one instant to another for an individual realization of the atmospheric turbulence. For such situations, numerically simulating the propagation of the beam is the preferred option. Simulations also provide a detailed beam profile as it propagates through the random medium, which is very useful for gauging the effect of the turbulence and the system performance. We want to further point out that most prior studies on beam propagation through the atmosphere have dealt with scalar Gaussian beams. Numerical simulation methods described in this chapter can be beneficial for understanding and visualizing arbitrary structured beams (e.g. beams carrying OAM or polarization singularities) through realistic turbulence levels

The beam propagation studies through the atmosphere usually undertake one of the two approaches—(a) the corpuscular [19] and (b) the wave approach [20]. In the corpuscular approach, the propagation of light is considered a random process of photon scattering by air molecules. This approach involves generating an ensemble of calculated photon trajectories, which can then be used to find the angular distribution, the polarization of the scattered beam, etc [19]. The second approach is the wave approach, which uses multiple phase screens to mimic the role of the atmosphere by perturbing the phase of the propagating wave [21–23]. Here, the beam propagation is considered a process of successive scattering of the light waves by these phase screens which have been generated by filtering Gaussian random noise with the atmospheric phase spectrum. This random phase screen method, also known as the split-step propagation method, is widely used in propagation studies [9, 24–29] of waves in various types of random media, particularly when one is interested in keeping track of the associated field quantities rather than just the total irradiance.

In what follows, we describe various concepts involved in numerical simulation of atmospheric laser beam propagation using the split-step propagation method. The detailed discussion should make it easier for the readers to follow (and possibly modify) the computer simulation code for split-step method in appendix A. The extended atmosphere is represented by stacking a finite number of thin phase screens along the propagation direction, separated by carefully chosen distance δz in free space depending on the turbulence strength. The essential details like the sampling criterion for the simulation grid, propagation path, number of phase screens to be used, etc are described. The chapter also discusses some recent methods used for generating random phase screens and their limitations.

8.2 Split-step propagation method

The split-step propagation method is a popular approach for modeling electromagnetic wave propagation through the atmosphere by numerically solving the parabolic wave equation [2]. This method provides a full-forward wave calculation with relatively few approximations and accommodates both vertical and lateral refractive index inhomogeneities in the atmosphere's refractive index [30–32]. This

Fourier-based split-step method belongs to the family of pseudo-spectral methods. It is widely used in computational mathematics to solve the time-dependent non-linear partial differential equations like the non-linear Schrödinger equation [33].

The parabolic wave equation method was first used to study the long-range propagation of radio waves in the troposphere in 1946 by Leontovich and Fock [34]. However, it was soon observed that this analytical method was not apt for dealing with complicated atmospheric conditions. It was only after Hardin and Tappert introduced an algorithm based on the Fourier split-screen method for numerically solving the parabolic equation that this method gained popularity in both acoustics and atmospheric wave propagation studies [35–37]. The split-step method discussed by Hardin and Tappert can be applied to differential equations with the property that the terms with the highest derivatives are linear with constant coefficients, and the non-linear or variable coefficient terms have fewer or no derivatives [35, 38]. The essence of this method is to alternate between the following two steps [35]: (i) advance the solution using only the non-linear or variable coefficient term using an implicit finite difference approximation and (ii) advance the solution exactly using only the linear, constant-coefficient term through the fast Fourier transform (FFT). The split-step formulation for the optical parabolic wave equation is described in the next section.

8.2.1 Split-step formulation from the parabolic equation

Consider the parabolic equation for wave propagation through the atmosphere that may be obtained by neglecting the second derivative of the field in z:

$$\nabla_T^2 U + 2k^2 n_1 U + 2ik\frac{\partial U}{\partial z} = 0. \tag{8.1}$$

Here $\nabla_T^2 = \frac{\partial^2}{\partial x^2} + \frac{\partial^2}{\partial y^2}$ represents the transverse Laplacian operator. This equation can be rearranged as,

$$\frac{\partial U}{\partial z} = \left[\frac{i}{2k}\nabla_T^2 + ikn_1\right]U \tag{8.2}$$

In order to apply the split-step formulation to this equation, it is necessary that the refractive index inhomogeneities of the medium satisfy the Markov assumption. In other words, the local index fluctuations $n_1(x, y, z)$ are delta-correlated (or un-correlated between two z-slices) along the wave propagation direction. Consider the field $U(x, y, z)$ as it propagates between two planes, one located at $z_1 = z_o$ and the other at $z_2 = z_o + \delta z$ where δz is chosen to be larger then the outer scale of the refractive index inhomogeneities but small enough to apply any perturbation method locally. It is then assumed that the field $U(x, y, z_2)$ can be calculated from the field $U(x, y, z_1)$ by adding small perturbations to this field locally [10]. This permits us to express the propagated field $U(x, y, z_2)$ in terms of $U(x, y, z_1)$ as:

$$U(x, y, z_o + \delta z) = \exp\left[\frac{i}{2k}\delta z\ \nabla_T^2 + ik\int_0^{\delta z} n_1(x, y, z)dz\right]U(x, y, z_o) \tag{8.3}$$

The first term in the exponent refers to effect of the transverse field derivatives while the second term refers to the effect of the refractive index fluctuations. By making use of the Markov approximation, these two terms in the exponent can be separated with negligible error [2]. Equation (8.3) can be approximated analytically with second-order accuracy by the split operator:

$$U(x, y, z_o + \delta z) = \exp\left[ik\int_0^{\delta z} n_1(x, y, z)dz\right]\exp\left[\frac{i}{2k}\delta z\nabla_T^2\right]U(x, y, z_o) \quad (8.4)$$

This is the basic philosophy behind the split-step propagation method, which allows one to numerically solve the parabolic wave equation. It can be seen from equation (8.4) that the propagation of field through the atmosphere is reduced to two independent steps:

1. Propagation through free space to a distance δz by using the propagator:

$$h_f(x, y, z) = \exp\left[\frac{i}{2k}\delta z\nabla_T^2\right] \quad (8.5)$$

2. Multiplying the resultant field by a phase function that represents the effect of the refractive index fluctuations of the medium over the same distance δz:

$$t(x, y) = \exp\left[i\theta(x, y)\right] = \exp\left[ik\int_0^{\delta z} n_1(x, y, z)dz\right] \quad (8.6)$$

   This step is equivalent to multiplying the field with a random phase screen with transmission function $t(x, y) = \exp[i\theta(x, y)]$. Therefore, the split-step method is also known as the random phase screen method, where the atmosphere is described by a finite number of thin phase screens stacked along the propagation direction and separated by equal distance δz in free space. These phase screens mimic the atmosphere by perturbing the phase of the propagating wave according to the atmospheric phase spectrum.

The first step of the split-step algorithm involves propagation of the field through free space by using the propagator given in equation (8.5). For this step, the fluctuating part of refractive index $n_1(x, y, z) = 0$, so that

$$\frac{\partial U}{\partial z} = \frac{i}{2k}\nabla_T^2 U \quad (8.7)$$

One can obtain the Fourier transform of this equation by using the following identities [2]:

$$\mathcal{F}\left[\frac{\partial g(x)}{\partial x}\right] = i2\pi f_x G(f_x) \quad (8.8)$$

$$\mathcal{F}\left[\frac{\partial^2 g(x)}{\partial x^2}\right] = -(2\pi f_x)^2 G(f_x) \quad (8.9)$$

where $G(f_x)$ is the Fourier transform of the function $g(x)$. This gives us in the Fourier-transform space,

$$\frac{\partial \tilde{U}}{\partial z} = i\pi\lambda(f_x^2 + f_y^2)\tilde{U} \tag{8.10}$$

where $\widetilde{U} = \mathcal{F}[U]$. This equation can be integrated over z to obtain the free-space propagation in the Fourier-transform space as,

$$\tilde{U}(f_x, f_y, z_o + \delta z) = H_f(f_x, f_y)\tilde{U}(f_x, f_y, z_o) \tag{8.11}$$

where $H_f(f_x, f_y) = \exp[i\pi\lambda\delta z(f_x^2 + f_y^2)]$ is the Fresnel transfer function. Therefore, using the transfer function notation, the field $U(x, y, z_o + \delta z)$ is given by [2, 9]

$$U(x, y, z_o + \delta z) = \exp\left[ik\int_0^{\delta z} n_1(x, y, z)dz\right]\mathcal{F}^{-1}\left[\exp\left(i\pi\lambda\delta z[f_x^2 + f_y^2]\right)\tilde{U}(f_x, f_y, z_o)\right] \tag{8.12}$$

The Fresnel transfer function $H_f(f_x, f_y)$ is commonly referred to as a 'chirp' function because the absolute value of its phase increases as the square of the spatial frequency variable [39, 40]. Such chirp functions are not band-limited and thus require careful choice of sampling conditions for aliasing-free simulation using FFT routines [23, 40, 41]. The phase screens used in atmospheric propagation simulations have their own set of sampling criterion, thus making it further challenging to do error-free Fresnel diffraction-based calculations. Though many algorithms have been developed to carry out fast Fresnel diffraction calculations [23, 42], in the current book, we refrain from using them. Instead we use the angular spectrum method for free-space propagation as explained thoroughly in chapter 3. Such an approach helps us in two ways: (a) we use the full Helmholtz wave equation for the free-space propagation, and (b) the sampling requirements are somewhat simpler. Therefore, the field $U(x, y, z_o + \delta z)$ using the angular spectrum method is now given by:

$$U(x, y, z_o + \delta z) = \exp\left[ik\int_0^{\delta z} n_1(x, y, z)dz\right]\mathcal{F}^{-1}\left[\exp\left[i\delta z\sqrt{k^2 - 4\pi^2(f_x{}^2 + f_y{}^2)}\right]\tilde{U}(f_x, f_y, z_o)\right] \tag{8.13}$$

The Fourier-based split-step algorithms provide an efficient way to numerically propagate the field through the atmosphere. The implementation of the split-step method is described in detail in the next section.

8.2.2 Implementation of the split-step propagation method

The geometry for wave propagation using the split-step method is illustrated in figure 8.2. The random phase screens are separated by distance δz and are placed at positions $(m - 1/2)\delta z$ with $m = 1, 2, 3, \ldots, M$ along the propagation direction where M is the total number of screens used. The phase screens carry the integrated phase due to turbulence over the distance δz. This random phase is obtained by integrating the refractive index fluctuation n_1 over the distance δz (see equation (8.6)). The optical field that is incident upon a random phase screen is multiplied with the transparency $t(x, y)$ of that screen and then propagated in free space to the

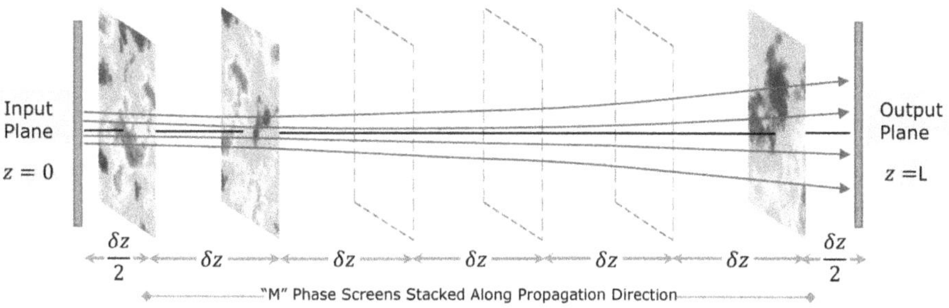

Figure 8.2. Schematic showing the placement of random phase screens in the split-step method for propagation of beams through atmospheric turbulence. Adapted with permission from [43] © 2019 The Optical Society of America.

next screen. The propagation is easily implemented using angular spectrum method. The two-dimensional (2D) Fourier transform of the field $U(x, y, z)$ is first multiplied with the free-space transfer function α for propagation over distance δz:

$$\alpha(f_x, f_y) = e^{i\delta z\sqrt{k^2 - 4\pi^2(f_x^2 + f_y^2)}}, \tag{8.14}$$

and the product is then inverse Fourier transformed to obtain the propagated field $U(x, y, z + \delta z)$. This propagated field now illuminates the next random phase screen and the above process continues till the final distance L is reached. The required steps of the propagation algorithm are shown in the flowchart of figure 8.3.

The accurate representation of the atmosphere and the free-space diffraction of the beam depends on various important factors like selecting the simulation grid points and minimization of boundary effects in FFT-based propagation routines, adequate sampling of the propagation path, and accuracy of phase reproduction on the random screens. The effect of these factors on the simulation accuracy can be minimized by fulfilling appropriate sampling requirements. The critical sampling criteria are described in detail in the following discussion.

8.2.3 Sampling requirements

Various sampling requirements need to be taken into account for accurate simulation of the wave propagation. These are described below:

1. **Simulation grid parameters**:

   In order to get rid of aliasing error in the FFT-based propagation method, it is necessary to choose the spatial grid sampling ($\Delta x = \Delta y$) and spatial frequency grid spacing ($\Delta f_x = \Delta f_y$) according to the Nyquist sampling requirements. Additionally, it may be required for an expanding beam to change the size of the computational window or sampling interval at different distances. Following the discussion in [40], the ideal sampling criterion for the free-space transfer function given in equation (8.14) is

$$\Delta x = \frac{\lambda\sqrt{(L^2 + (D/2)^2)}}{D} \tag{8.15}$$

Set-up Propagation Conditions:

- **Initial Beam** : Beam Type - $U_0(\boldsymbol{r}; z = 0)$, Beam Waist (W_o), Wavelength (λ)

- **Turbulence Condition** : Phase Spectrum Type – Von Karman $(\varphi_\theta(\kappa))$, C_n^2, Inner (l_o) and Outer (L_o) scales of Turbulence , Number of Screens (M)

- **Propagation Geometry** : Number of grid points (N), spatial sampling $\left(\Delta x = \Delta y \leq \frac{l_o}{2}\right)$, spatial frequency sampling $\left(\Delta f_x = \Delta f_y = \frac{1}{N\Delta x}\right)$, total Propagation distance (L), Inter-screen distance $\left(\delta z = \frac{L}{M}\right)$

Propagation to distance $\delta z/2$: Angular Spectrum Method

$$U_1(\boldsymbol{r}', z + \delta z/2) = FFT^{-1}\{FFT[U_0(\boldsymbol{r}, z)]\, e^{i\frac{\delta z}{2}\sqrt{k^2 - 4\pi^2(f_x^2 + f_y^2)}}\}$$

Generate a random phase screen $\theta(x', y')$:
Using Von Karman phase spectrum:

$$\varphi_\theta(\kappa) = 0.033 C_n^2 (\kappa^2 + \kappa_0^2)^{-11/6} e^{-\frac{\kappa^2}{\kappa_m^2}}$$

where $\kappa_o = \frac{2\pi}{L_0}$ and $\kappa_m = \frac{5.92}{l_o}$

Multiply the field with a generated random phase screen:

$$U_2(\boldsymbol{r}', z + \delta z/2) = U_1(\boldsymbol{r}', z + \delta z/2) \times e^{i\theta(\boldsymbol{r}')}$$

Propagation to distance $\delta z/2$: Angular Spectrum Method

$$U_3(\boldsymbol{r}'', z + \delta z) = FFT^{-1}\{FFT[U_2\left(\boldsymbol{r}', z + \frac{\delta z}{2}\right)]e^{i\frac{\delta z}{2}\sqrt{k^2 - 4\pi^2(f_x^2 + f_y^2)}}\}$$

Is the final distance L reached ?

No

Yes

Putting :
$$U_o(\boldsymbol{r}, z) = U_3(\boldsymbol{r}'', z + \delta z)$$

Analyse Output

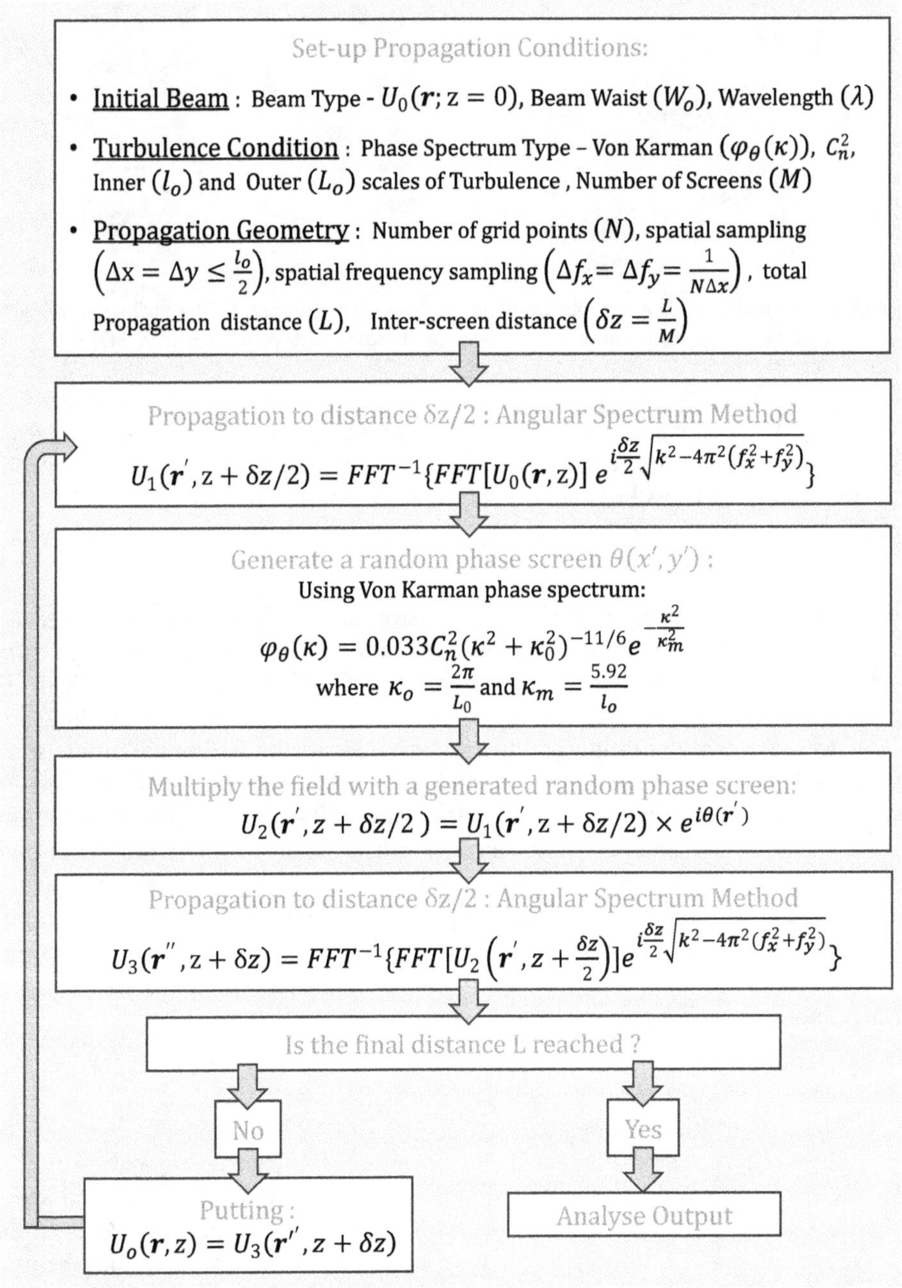

Figure 8.3. Flowchart denoting the steps for numerically propagating the beam through atmospheric turbulence using the random phase screen method.

where L is the total propagation length and $D = N\Delta x$ is the total spatial size of the simulation grid. Here N denotes the number of grid points. For paraxially propagating beams or when $L \gg D$, the sampling criterion simplifies to,

$$\Delta x = \frac{\lambda L}{D} = \sqrt{\frac{\lambda L}{N}} \tag{8.16}$$

From the discussion of the free-space transfer function α in chapter 3, we know that spatial frequencies higher than $1/\lambda$ cannot propagate in free space. By choosing the spatial grid sampling period $\Delta x > \lambda/2$, the maximum allowed spatial frequency on the grid becomes: $f_{\max} = \frac{1}{2\Delta x} < \frac{1}{\lambda}$. This ensures that one is only dealing with propagating spatial frequencies.

In case the computation needs to be performed with a larger sampling grid spacing $\Delta x'(=\Delta y')$, a low-pass filter defined as:

$$LP(f_x, f_y) = 1, \quad f_x^2 + f_y^2 < = f_0^2 \tag{8.17}$$

and zero elsewhere, needs to be used additionally for aliasing-free propagation computation [44]. Here the quantity f_0 representing the radius of the low-pass filter is given as:

$$f_0 = \frac{1}{\lambda\sqrt{1 + (2z/N\Delta x')^2}}. \tag{8.18}$$

2. **Accurate dynamic range**:

   One important limitation in the applicability of numerical simulations is the finite spatial dynamic range imposed by the maximum available numerical grid size. For a given simulation grid, the maximum and minimum allowed wave numbers are $\kappa^G_{\max} = \frac{\pi}{\Delta x}$ and $\kappa^G_{\min} = \frac{\pi}{N\Delta x}$, respectively, with the available dynamic range,

$$\frac{\kappa^G_{\max}}{\kappa^G_{\min}} = N \tag{8.19}$$

The spectral range defined by the numerical simulation parameters $[\kappa^G_{\max}, \kappa^G_{\min}]$ must provide sufficient bandwidth to include all the essential components of the spatial spectrum of the intensity fluctuations which are produced on propagation through turbulence. In this regard, the asymptotic solutions for spherical wave propagation in turbulence can be used to provide the required range of correlation scales of the intensity fluctuations. Though the asymptotic theory is valid only for strong path-integrated turbulence, it is useful in defining the extreme limits to the simulation spectrum requirements.

Two important correlation scales can be defined for the intensity fluctuations, d_o and D_o, as shown in figure 8.4. These two scales d_o and D_o are associated with the diffractive and refractive effects of the atmospheric

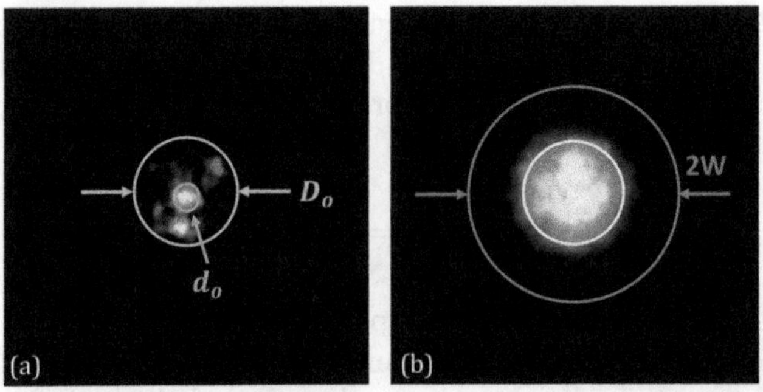

Figure 8.4. The short-time (instantaneous) and long-time averaged beam spot of a initially collimated Gaussian beam of $1.55\mu m$ wavelength are shown in (a) and (b). The beam has been propagated over 2 km through moderate atmospheric turbulence given by $C_n^2 = 10^{-14}\ \mathrm{m}^{-2/3}$. Here, the diffractive short scale and refractive long scale are denoted by d_o and D_o respectively. The white circle shows the corresponding free space beam spot diameter.

turbulent eddies, respectively. The atmosphere contains turbulent eddies of different sizes ranging from the outer scale L_o to the inner scale l_o. These eddies depending on their spatial scales have different effects on the propagating wave. The larger-scale eddies produce *refractive* effects which cause small-angle scattering and focusing resulting in beam wandering, whereas the smaller-scale eddies produce *diffractive* effects which produce amplitude fluctuations and beam spreading. The small-scale $d_o \simeq r_{os}$ describes the feature size of the speckle pattern superimposed on the beam's irradiance profile. Here r_{os} denotes the Fried parameter for spherical waves given by the expression,

$$r_{os} = \left[0.42k^2 \int_0^L C_n^2(z)\left(\frac{z}{L}\right)^{5/3} dz\right]^{-3/5} \tag{8.20}$$

The larger-scale D_o originates in the strong scattering regime and is known as the effective beam spreading diameter. In isotropic turbulence, the scale D_o has approximately the width of the scattering disk size, that is, $D_o = 4L/kr_{os}$. The intensity is therefore modeled as a diffractive process of scale d_o modulated by a larger refractive process of scale D_o. The intensity spectrum is thus band-limited by a minimum $\kappa_{\min}^S = \pi/D_o$ and maximum spatial wave number $\kappa_{\max}^S = \pi/d_o$. The required dynamic range for propagation simulation of spherical waves through turbulence is equal to

$$\frac{\kappa_{\max}^S}{\kappa_{\min}^S} = \frac{D_o}{d_o} = \frac{4L}{k{r_{os}}^2} \tag{8.21}$$

For proper representation of all the required spatial frequencies in the simulation,

$$\frac{\kappa_{\max}^{G}}{\kappa_{\min}^{G}} \geqslant n\frac{\kappa_{\max}^{S}}{\kappa_{\min}^{S}} \tag{8.22}$$

where n is an integer $\geqslant 4$ for safely avoiding any aliasing errors. The necessary sampling condition from equations (8.19) and (8.20) can be stated as:

$$N \geqslant n\frac{4L}{kr_{os}^{2}}. \tag{8.23}$$

3. **Beam Broadening considerations**:
   Consider a Gaussian beam with its field in $z = 0$ plane given by:

$$\psi(\boldsymbol{r};\, z = 0) = A\exp\left(-\frac{r^2}{W_o^2} - \frac{i\pi r^2}{\lambda F_o}\right) \tag{8.24}$$

where A and $F_o(>0)$ denote the complex amplitude and the radius of curvature of the wavefront. Here, W_o denotes the beam waist (or the $1/e^2$ beam intensity radius) at the $z = 0$. In free space, the beam waist at a distance $z > 0$ can be computed using Fresnel propagation and is given by the expression:

$$W(z) = W_o\sqrt{\left(1 - \frac{z}{F_o}\right)^2 + \left(\frac{z\lambda}{\pi W_o^2}\right)^2} \tag{8.25}$$

However, in a turbulent medium, the propagating wave not only expands due to diffraction but also experiences a turbulence-induced beam spreading. Turbulent eddies larger than the beam diameter deflect the beam spot and on propagation over larger distances, this deflection can become a serious issue as now the beam no longer falls onto the detector. The instantaneous beam width is given by the short-term beam radius defined as $W_s(z)$. The long-time averaged beam radius is given by:

$$W^2(z) = W_s^2(z) + 2\langle\beta^2(z)\rangle \tag{8.26}$$

where $\beta(z)$ denotes the beam wander defined by the position of the beam centroid. Beam wander establishes the difference between the short- and long-term beam radius and is usually given as,

$$\beta(z) = [\beta_x(z),\, \beta_y(z)] = \frac{\iint \rho U(\rho,\, z)d\rho}{\iint U(\rho,\, z)d\rho} \tag{8.27}$$

and

$$\langle\beta(z)^2\rangle = \langle\beta_x^2\rangle + \langle\beta_y^2\rangle \tag{8.28}$$

Here, $U(\rho, z) = U(x, y, z)$ is the instantaneous beam intensity profile in the plane z. The short- and long-time averaged beam spots are shown in figure 8.4.

It is necessary to consider the relation between the beam size and the lateral dimension of the simulation grid ($N\Delta x$) so that the broadened beam is properly contained inside the simulation grid area for all propagation distances of interest. An approximate relation for the long-time averaged beam radius in a turbulent medium was obtained by Yura [45] using the extended Huygens–Fresnel principle as:

$$W^2(z) = W_o^2\left[\left(1 - \frac{z}{F_o}\right)^2 + \left(\frac{\lambda z}{\pi W_o^2}\right)^2\right] + \left(\frac{D_o}{2}\right)^2, \tag{8.29}$$

The spatial extent of the simulation grid should be chosen such that

$$N\Delta x \geqslant 2\,mW(z) \tag{8.30}$$

where m is an integer. The power contained in the tail of the Gaussian beam can be neglected outside a circular area of diameter $2W(z)$. Therefore, by choosing $m \geqslant 3$, it is ensured that the beam is contained within the simulation grid as it propagates, and no edge effects creep up. The other means of achieving this is by imposing absorption boundaries or using windowed illumination fields like super-Gaussian beams.

4. **Accurate phase representation and sampling of turbulence parameters**:

The phase screens used in the split-step propagation method also need to fulfill some numerical requirements for representing the atmospheric phase fluctuations accurately. The statistics of the beam intensity are susceptible to the small-scale phase fluctuations. So an important criterion for phase screens in this respect is that the inner scale of turbulence should be sampled appropriately. This is ensured by choosing $\Delta x \leqslant l_o/2$ according to the Nyquist sampling. Besides this, the change in the phase of the phase screen from one grid point to the other should be less than π again to fulfill Nyquist sampling,

$$\theta(x_2, y_2) - \theta(x_1, y_1) < \pi \tag{8.31}$$

or in terms of $\Delta x = x_2 - x_1$ and $\Delta y = y_2 - y_1$, we have

$$\Delta x\left|\frac{\partial\theta(x, y)}{\partial x}\right| < \pi, \tag{8.32}$$

and

$$\Delta y\left|\frac{\partial\theta(x, y)}{\partial y}\right| < \pi \tag{8.33}$$

5. **Number of random phase screens required for given turbulence strength**:
   The propagation path L also needs to be sufficiently sampled in the z direction. This is usually achieved by following the thumb rule that less than 10% of the total intensity scintillation takes place over the inter-screen distance δz, that is, $\sigma_I{}^2(\delta z) < 0.1\ \sigma_I{}^2(L)$, where $\sigma_I{}^2(z)$ gives the intensity variance at any z plane. The Rytov variance (σ_R^2) is used as an estimate for intensity variance. It is further required that the intensity scintillation is weak over the distance δz. This can be ensured by placing a requirement that $\sigma_I{}^2(\delta z) < 0.1$. These two conditions can be used for determining the minimum number of screens required for any given value of $C_n{}^2$ and their placement along the propagation path. An estimate for obtaining the minimum number of phase screens has been provided in [46]:

$$M > (10\sigma_R^2)^{6/11} \tag{8.34}$$

   The phase screens should also be delta-correlated along the z direction as per the Markov approximation. For this, δz must be chosen to be higher than the outer scale of turbulence L_o.

All the above conditions are essential for accurate modeling of the atmospheric wave propagation. In the next section, we will study the various methods for generating random phase screens in detail.

8.3 Phase screen generation

The simplest random phase screen model made use of only one phase screen and was first used in the analysis of wave diffraction in a thick slab of random inhomogeneous medium [47]. Such a treatment is useful when considering beam propagation through ionosphere or interplanetary plasma [48]. However, a single phase screen is not sufficient to describe the propagation of beams through extended turbulence, for example, the beam propagation over a horizontal propagation path over a few km in atmosphere. The error arising due to the representation of a continuum random medium by a single phase screen was studied by Booker *et al* [49]. It was observed that when the propagation length exceeds the outer scale ($z > L_o$), the single phase screen model significantly overestimates intensity fluctuations. In this case, a multiple phase screen model offers a more reliable representation of the turbulent medium [50].

The ability to generate accurate random phase screens lies at the core of the simulation for studying atmospheric turbulence effects. In order to simulate realistic turbulence phase fluctuations, the ensemble statistics of the random phase screens should match with those predicted by the theory. A large number of methods have been developed for phase screen generation. These methods can be broadly classified into two main types. The first type of phase screen generation method makes use of the Fourier transform to represent the phase screen as a 2D rectangular grid of points by filtering complex Gaussian random numbers with the desired power spectral density of the phase fluctuations. The second type of phase screen generation method represents the phase screen as a sum of orthogonal basis functions [18, 51, 52]. In this section, we will concern ourselves with the FFT-based phase screen generation method.

The atmospheric phase is a random function of the refractive index fluctuations. The atmospheric phase spectrum $\Phi_\theta(\kappa)$ corresponding to the turbulence-induced phase fluctuations can be obtained from the power spectrum of refractive index fluctuations $\Phi_n(\kappa)$ as described in the following section.

8.3.1 Phase spectrum from the refractive index spectrum

Consider a thin slab of random medium of thickness δz whose refractive index fluctuations are given by $n_1(\boldsymbol{r}, z)$. From equation (8.6), the 2D phase fluctuations $\theta(\boldsymbol{r})$ on passing through this slab are given by:

$$\theta(\boldsymbol{r}) = k\int_0^{\delta z} n_1(\boldsymbol{r}, z)dz \tag{8.35}$$

The correlation function for phase can thus be written as:

$$\begin{aligned}\Gamma_\theta(\boldsymbol{r}_1, z_1; \boldsymbol{r}_2, z_2) &= \langle\theta(\boldsymbol{r}_1, z_1)\theta(\boldsymbol{r}_2, z_2)\rangle \\ &= k^2\iint_0^{\delta z}\langle n_1(\boldsymbol{r}_1, z_1)n_1(\boldsymbol{r}_2, z_2)\rangle dz_1dz_2\end{aligned} \tag{8.36}$$

where $\langle\ \rangle$ denotes ensemble average. The quantity $\langle n_1(\boldsymbol{r}_1, z_1)n_1(\boldsymbol{r}_2, z_2)\rangle$ is the correlation function $\Gamma_n(\boldsymbol{r}_1, z_1; \boldsymbol{r}_2, z_2)$ of the refractive index fluctuations. Under the assumption that δz is much larger than the correlation length of the refractive index irregularities, one can make use of the Markov approximation to write $\Gamma_n(\boldsymbol{r}_1, z_1; \boldsymbol{r}_2, z_2)$ as

$$\Gamma_n(\boldsymbol{r}_1, z_1; \boldsymbol{r}_2, z_2) = \langle n_1(\boldsymbol{r}_1, z_1)n_1(\boldsymbol{r}_2, z_2)\rangle = \delta(z_1 - z_2)F_n(\boldsymbol{r}_1 - \boldsymbol{r}_2), \tag{8.37}$$

where $F_n(\boldsymbol{r}_1 - \boldsymbol{r}_2)$ is the 2D power spectrum of the refractive index fluctuations in the transverse x–y plane. The function $F_n(\boldsymbol{r}_1 - \boldsymbol{r}_2)$ may be obtained from $\Phi_n(\kappa)$ by integrating over the z plane, that is,

$$F_n(\boldsymbol{r}) = 2\pi\iint_{-\infty}^{+\infty}\Phi_n(\kappa_x, \kappa_y, \kappa_z = 0)\exp(i\boldsymbol{\kappa}\cdot\boldsymbol{r})d\kappa_xd\kappa_y, \tag{8.38}$$

where $\boldsymbol{r} = \boldsymbol{r}_1 - \boldsymbol{r}_2$. Substituting equation (8.37) in equation (8.36), one can write,

$$\Gamma_\theta(\boldsymbol{r}_1, z_1; \boldsymbol{r}_2, z_2) = k^2\iint_0^{\delta z}\delta(z_1 - z_2)F_n(\boldsymbol{r})dz_1dz_2 = k^2\delta z\, F_n(\boldsymbol{r}). \tag{8.39}$$

Next, we define a power spectral density $\Phi_\theta(\boldsymbol{\kappa})$ associated with the random phase fluctuations, which forms a Fourier transform pair (Wiener–Khintchine theorem) with the phase correlation function $\Gamma_\theta(\boldsymbol{r})$, such that

$$\Phi_\theta(\boldsymbol{\kappa}) = \int d\boldsymbol{r}\ \Gamma_\theta(\boldsymbol{r})\exp(-i\boldsymbol{\kappa}\cdot\boldsymbol{r}) \tag{8.40}$$

Now, using equations (8.39) and (8.40), we see that the phase spectrum $\Phi_\theta(\kappa)$ takes the form,

$$\Phi_\theta(\kappa) = 2\pi\ k^2\ \delta z\ \Phi_n(\kappa_x,\ \kappa_y,\ \kappa_z = 0) \tag{8.41}$$

This expression allows one to obtain the phase spectrum corresponding to a given refractive index spectrum $\Phi_n(\kappa)$.

The refractive index fluctuations can be modeled as a stochastic process. The phase $\theta(\boldsymbol{r})$ is essentially a sum over many n_1 values (see equation (8.35)) as δz is taken to be larger than the correlation length of the turbulence. This enables us to model the phase $\theta(\boldsymbol{r})$ as a Gaussian random process as per the central limit theorem. As explained in chapter 2, the traditional method for generating realizations of a random field with a well-defined power spectrum involves filtering of Gaussian white noise with the square root of the desired spectrum, followed by an inverse Fourier transform operation. This method is known as the FFT method and is commonly used to generate random realizations of atmospheric turbulence in simulation studies. In the next section, we will describe the FFT method for phase screen generation in detail. A few other known methods used for random screen generation will also be briefly discussed in the end.

8.3.2 FFT method for phase screen generation

The topic of generating realizations of a random process with a given power spectrum was already discussed briefly in section 2.3. A continuous 2D random phase screen $\theta(\boldsymbol{r})$ can be represented as a Fourier integral

$$\theta(\boldsymbol{r}) = \iint_{-\infty}^{+\infty} g(\kappa_x,\ \kappa_y)\sqrt{\Phi_\theta(\kappa_x,\ \kappa_y)}\exp(i\boldsymbol{\kappa}\cdot\boldsymbol{r})d\kappa_x d\kappa_y \tag{8.42}$$

where $\boldsymbol{\kappa} = [\kappa_x,\ \kappa_y,\ \kappa_z = 0]$ and $g(\kappa_x,\ \kappa_y)$ is a zero mean, unit-variance complex Gaussian white noise process. In order to simulate the phase screen on a numerical grid, equation (8.42) needs to be converted to its approximate discrete form. The discrete phase screen is written as a sum of Fourier harmonics with random complex coefficients,

$$\theta_{\mathrm{FFT}}(x,\ y) = \sum_{n,\ m=-N/2}^{N/2-1} g(\kappa_x,\ \kappa_y)\sqrt{\Phi_\theta(\kappa_x,\ \kappa_y)}\ \exp\left[i2\pi(f_x x + f_y y)\right]\ \Delta\kappa_x\ \Delta\kappa_y \tag{8.43}$$

This equation can be easily implemented using the numerically efficient FFT algorithm. In the above equation, $\kappa_x = 2\pi f_x$ and $\kappa_y = 2\pi f_y$. The sample points on the numerical grid in the spatial and frequency domain can be defined as $[x,\ y] = [j\Delta x,\ l\Delta y]$ and $[f_x, f_y] = [n\Delta f_x,\ m\Delta f_y]$, respectively. The sample intervals are chosen as $\Delta x = \Delta y$ and $\Delta f_x = \Delta f_y = 1/(N\Delta x)$ where N is the total number of grid points. The function $g(\kappa_x,\ \kappa_y)$ denotes a discrete complex Gaussian noise process given by:

$$g(\kappa_x,\ \kappa_y) = \frac{g(n,\ m)}{\sqrt{\Delta\kappa_x\ \Delta\kappa_y}} = \frac{a(n,\ m) + ib(n,\ m)}{\sqrt{\Delta\kappa_x\ \Delta\kappa_y}} \tag{8.44}$$

where the discrete random numbers $a(n, m)$ and $b(n, m)$ are drawn from a Gaussian random distribution with standard deviation equal to unity. Therefore, the expression for the simulated phase screen is given by:

$$\theta_{\text{FFT}}(j\Delta x, l\Delta y) = \sum_{n,\, m\, =-N/2}^{N/2-1} g(n, m)\sqrt{\Phi_\theta(n\Delta\kappa, m\Delta\kappa)}\ \exp\left[i\frac{2\pi}{N}(jn + lm)\right]\ \Delta\kappa \quad (8.45)$$

The FFT operation above is expected to give a complex-valued function in general whose real part is typically used as the required phase function θ_{FFT}. The value of the phase spectrum Φ_θ at the origin ($n = m = 0$) is usually set equal to zero. This has no effect on the spatial statistics of the wave field as the zero-frequency component of the phase-screen only determines the average phase delay offered by the screen. This term has no contribution to the tip, tilt or behavior of the propagating field. Besides, it has been shown that a large enough value of $\Phi_\theta(0, 0)$ may lead to a quantization error [53]. Figure 8.5 details the steps for the random screen generation using the FFT method. The generated sample phase screens for two different C_n^2 values: $C_n^2 = 10^{-13}$ m$^{-2/3}$ and $C_n^2 = 10^{-15}$ m$^{-2/3}$, respectively, are also shown in figures 8.5 (a) and (b). Note that the range of phase variation in the phase screens for a given δz

FFT Method : Phase Screen Generation

Generate a $(N \times N)$ array of complex Gaussian random numbers : $g(n, m) = a(n, m) + ib(n, m)$

Multiply with the square root of the phase spectrum : $\Delta\kappa\sqrt{\Phi_\theta(n\Delta\kappa, m\Delta\kappa)}$
where $\Delta\kappa = 2\pi\Delta f_x$ and put $\Phi_\theta(\kappa = 0) = 0$

Inverse FFT to get random screen
$$\theta_{FFT}(j\Delta x, l\Delta y) = \sum_{n,\, m=-N/2}^{N/2-1} g(n, m)\sqrt{\Phi_\theta(n\Delta\kappa, m\Delta\kappa)}\, e^{\frac{i2\pi}{N}(jn+lm)}\, \Delta\kappa = \theta_1 + i\theta_2$$

Pick θ_1 as phase screen

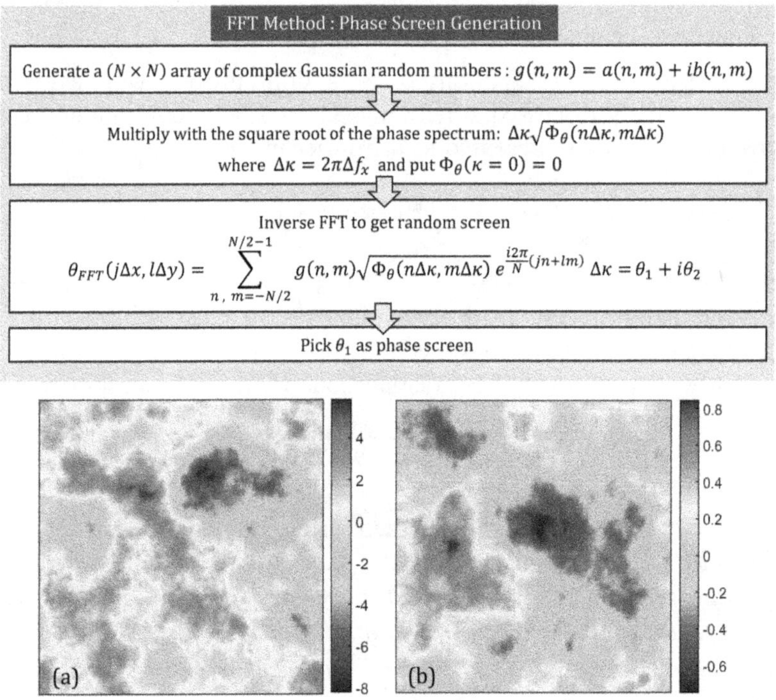

Figure 8.5. The flowchart depicts the steps of the FFT method for random phase screen generation. Here (a) and (b) show examples of generated phase screens when the structure function for the refractive index fluctuations $C_n^2 = 10^{-13}$ m$^{-2/3}$ and $C_n^2 = 10^{-15}$ m$^{-2/3}$, respectively. The above phase screens are generated using a 512×512 simulation grid with von Kármán phase spectrum having inner and outer scales equal to $l_o = 1$ cm, $L_o = 10$ m, respectively and the propagation step length is taken as $\delta z = 25$ m.

is dependent on the value of the structure constant C_n^2 of the refractive index fluctuations.

Accuracy of the FFT-based random screen simulator

The accuracy of the phase screens generated by any method is typically evaluated by their ability to reproduce the desired phase structure function for the given turbulence model. The structure function for the phase fluctuations is defined as,

$$D_\theta(\boldsymbol{r}) = \langle [\theta(\boldsymbol{r}_1) - \theta(\boldsymbol{r}_2)]^2 \rangle, \tag{8.46}$$

which represents the average squared difference in the phase of the screen for pair of points which are separated by $\boldsymbol{r} = \boldsymbol{r}_1 - \boldsymbol{r}_2$ from each other. The structure function is related to the 2D phase correlation function $\Gamma_\theta(\boldsymbol{r})$ of the phase screen as,

$$D_\theta(\boldsymbol{r}) = 2[\Gamma_\theta(0) - \Gamma_\theta(\boldsymbol{r})] \tag{8.47}$$

where

$$\Gamma_\theta(\boldsymbol{r}) = \langle \theta(\boldsymbol{r}_1)\theta(\boldsymbol{r}_2) \rangle \tag{8.48}$$

The 2D phase correlation function is equal to the Fourier transform of the 2D phase spectrum,

$$\Gamma_\theta(\boldsymbol{r}) = \iint_{-\infty}^{+\infty} \Phi_\theta(\kappa_x, \kappa_y, \kappa_z = 0)\exp(i\boldsymbol{\kappa} \cdot \boldsymbol{r})d\kappa_x d\kappa_y \tag{8.49}$$

Therefore, a discrete form of 2D correlation function for the FFT-based phase screen can be obtained by taking an inverse FFT of the discrete 2D phase spectrum[1] as,

$$\Gamma_{\theta_{\mathrm{FFT}}}(j\Delta x, l\Delta y) = \sum_{n,\, m = -N/2}^{N/2-1} \Phi_\theta(n\Delta\kappa, m\Delta\kappa)\ (\Delta\kappa)^2\ \exp\left[i\frac{2\pi}{N}(jn + lm)\right] \tag{8.50}$$

The above expression for the 2D correlation function can be substituted into equation (8.47) to calculate the expected structure function. It is observed that for a statistically large enough ensemble of random phase screens (>1000), the computed average structure function closely matches the expected structure function. The expected phase structure function for the random screens is compared with the theoretical structure function for the same atmospheric model and turbulence parameters. In general, the theoretical structure function for phase screens is given by

$$D_\theta(\boldsymbol{r}) = 2\iint_{-\infty}^{+\infty} \Phi_\theta(\boldsymbol{\kappa})\ [1 - \exp(i\boldsymbol{\kappa} \cdot \boldsymbol{r})]\ d\kappa_x\ d\kappa_y \tag{8.51}$$

For the case of modified von Kaŕmán turbulence spectrum, the theoretical phase structure function D_θ has the form:

[1] The pole at the origin of the phase spectrum is usually handled by putting the numerical spectrum value equal to zero.

$$D_\theta^{mvK}(r) = 3.08\ r_{of}^{-5/3}\left\{\Gamma\left(-\frac{5}{6}\right)\kappa_m^{-5/3}\left[1 - {}_1F_1\left(-\frac{5}{6};\ 1;\ -\frac{\kappa_m^2 r^2}{4}\right)\right] - \frac{9}{5}\kappa_0^{1/3} r^2\right\} \tag{8.52}$$

where ${}_1F_1(a;\ c;\ z)$ is a confluent hypergeometric function of the first kind and r_{of} represents the Fried parameter for plane wave,

$$r_{of} = \left[0.423k^2 \int_0^L C_n^2(z)dz\right]^{-3/5} \tag{8.53}$$

This structure function expression can be written in a simpler form within <2% error margin by using an algebraic approximation for the hypergeometric function, which yields

$$D_\theta^{mvK}(r) \approx 7.75\ r_{of}^{-5/3} l_o^{-1/3} r^2 \left[\frac{1}{\left(1 + \frac{2.03r^2}{l_o^2}\right)^{1/6}} - 0.72(\kappa_o l_o)^{1/3}\right] \tag{8.54}$$

Figure 8.6 compares the expected structure function for the FFT-based phase screens with the theoretical structure function. The plots in figures 8.6(a) and (b) are generated using the von Kármán phase spectrum with two different values of the outer scale $L_o = 10$ m and $L_o = 100$ m, respectively. The other simulation parameters are: $N = 512$, $D = 2$ m, $l_o = 1$ cm, $C_n^2 = 10^{-14}$ m$^{-2/3}$ and $\delta z = 100$ m. The difference between the theoretical structure function and the simulated structure function is readily apparent in the two plots. This big discrepancy in the behavior of the simulated structure function is due to the inadequate sampling of the low-frequency content of the 2D phase spectrum by the FFT-based simulation method.

The FFT-based simulation method is thus seen to suffer from the basic limitation that the available bandwidth of spectral components generated in the phase screen is

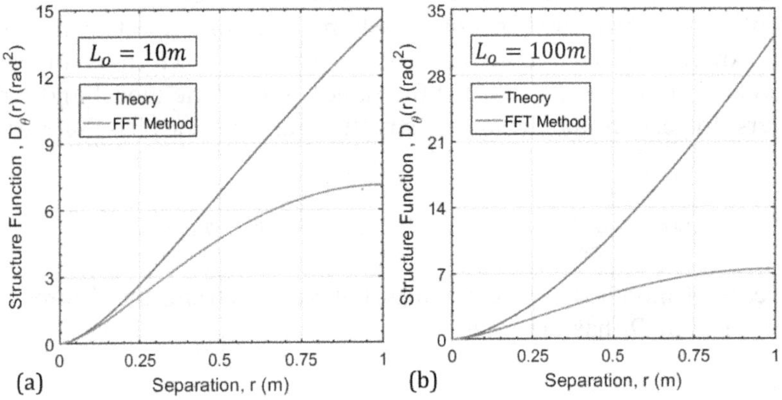

Figure 8.6. The structure function for the FFT-based phase screens is compared with the theoretical structure function. The simulation parameters are detailed in text.

decided by the numerical grid parameters. From the perspective of modeling atmospheric turbulence, this means that the width of the phase screen (D) decides the simulated outer turbulence scale while the sampling size (Δx) decides the simulated inner turbulence scale of the phase screens. It is quite possible that the parameters D and Δx associated with the phase screen would not accurately represent the actual inner and outer scales of the atmospheric turbulence. The maximum and minimum spatial frequencies of the simulation are given by $f_{\min}^{\text{sim}} = \Delta f = 1/D$ and $f_{\max}^{\text{sim}} = N\Delta f/2 = N/2D$. As the frequencies lower than $f_{\max}^{\text{sim}}$ are neglected, this means that the phase screen does not incorporate the effects due to the refractive index fluctuations having sizes greater than the screen size.

The inner scale is not represented properly on the numerical grid (or the phase spectrum is under-sampled at high frequencies) when the sampling size Δx is greater than the inner scale value l_o. Such cases are generally not severe due to the spectral roll-off at higher κ values (note the exponential decay term in the von Kármán spectrum model). However, under-sampling of the outer scale puts a serious limit on the applicability and accuracy of the FFT-based phase screens. The outer scale is the main contributor to the turbulence's low-frequency characteristics, for example, tilt. These lower frequencies are responsible for causing the wandering of the beam centroid, which may become significant for long propagation distances. The minimum and maximum spatial frequencies corresponding to the parameters used for simulating figure 8.6 are $f_{\min}^{\text{sim}} = 0.5\text{m}^{-1}$ and $f_{\max}^{\text{sim}} = 128\text{m}^{-1}$ which, respectively, corresponds to scale sizes of 2 m and 0.0078 m. Thus, the outer scale parameter $L_o = 10$ m in figure 8.6(a) and $L_o = 100$ m in figure 8.6(b) is not appropriately represented by the simulated screen. In order to properly sample the outer scale $L_o = 10$ m, the first sample away from origin in 2D Fourier space should occur at $f_{\min} = 1/L_o = 0.1\text{m}^{-1}$ whereas in the simulated screen, the first sample is located at $f_{\min}^{\text{sim}} = 0.5\text{m}^{-1}$. Therefore, a major portion of the spectral energy near the origin is not sampled appropriately and the resultant phase screens therefore do not have accurate large-scale turbulence characteristics.

A straightforward way to incorporate lower spatial frequencies in the phase screen is by increasing the size of the grid by increasing N and keeping Δx the same. However, this will result in very large grid sizes that are not practically feasible. For example, a grid size, which is five times the outer scale value, has been suggested for adequately sampling the outer scale [54]. In that case, for $L_o = 10$ m, one would require $D = 50$ m or an impractically high $N = 12\,800$. Another scheme that has been explored in this regard is to cut out a small portion from a very large phase screen and use it in propagation simulations [26]. It is expected that the lower frequency characteristics will be accurate over a small portion of a large screen. A Zernike polynomial-based representation of Kolmogorov's turbulence spectra was provided in [55]. One can add low-order Zernike modes using Noll's method to compensate for the low frequencies. Recently, low-frequency compensation using correlation matrix based methods have also been proposed [56]. While the methodology described above leads to a square-shaped phase screen, there are applications like laser beam tracking, where beam propagation through turbulence over a range of continuously varying directions is required. A phase screen that extends over large distances in one dimension then becomes desirable for simulation purposes. One straightforward approach for generating screens of this nature involves using rectangular windows in the FFT screen method where the spatial frequency sampling in two orthogonal directions needs to be adjusted appropriately. A computationally efficient 'infinite' random screen approach [57] has also been developed for cases when it is required to simulate a very long rectangular screen with slowly varying turbulence statistics. We will not describe this methodology in further detail in this book.

In the next section, we discuss a more commonly used method for accounting for these lower frequencies. This method involves generation and addition of a sub-harmonic phase screen to the existing FFT-based phase screen. Multiple methods have been proposed to incorporate the sub-harmonic frequencies in a phase screen [26, 54].

8.3.3 Sub-harmonic correction to the phase screen

An efficient way of adding low-frequency information without employing a very large-sized phase screen is through the sub-harmonic method [26, 54, 58–60]. The basic idea is to generate a separate low-frequency (sub-harmonic) phase screen using the FFT method that finely samples only the low-frequency portion of the phase spectrum. This sub-harmonic screen is then added to the phase screen (simulated as explained in section 8.3.2) for accurately modeling the large-scale atmospheric turbulence effects. A sub-harmonic is basically a sinusoidal function with a period larger than the screen size which has a spatial frequency f^{sub} much lower than the lowest sampled frequency $f_{\text{min}}^{\text{sim}}$ of the phase screen. The sub-harmonic screen is generated by sampling the phase spectrum in the vicinity of $\kappa = 0$, at a much smaller sampling interval f^{sub}, than the rest of the spectrum. In this section, we will discuss the generation of this sub-harmonic phase screen as presented by Lane, Glindemann, and Dainty [58], which was later modified by Frehlich [59] to improve convergence.

The central pixel corresponding to spatial frequencies $f_x = 0$ and $f_y = 0$ of the high frequency phase screen is divided into nine equally sized sub-pixels, as shown in figure 8.7(a). Note that each sub-pixel is 1/9 of the area of the original pixel size (Δf_x) and the sub-harmonic sample width is $\Delta f_x^{\text{sub}} = \Delta f_x/3$. This describes the first sub-harmonic level. Sample points are now placed in the eight outer sub-pixels which are then used to generate the low-frequency screen using the FFT method. The process is now repeated again on the central small sub-pixel at the origin, creating multiple sub-harmonic levels as needed. The sample sizes of the sub-pixels at the pth sub-harmonic level ($p \geqslant 1$) are given by $\Delta f_{xp}^{\text{sub}} = \Delta f_x/3^p$. The phase contributions from all the sub-harmonic levels are added together to generate the total sub-harmonic screen. Assuming N_p sub-harmonic levels, the sub-harmonic phase screen is given by:

$$\theta_{\text{SH}}(j\Delta x, l\Delta y) = \sum_{p=1}^{Np} \sum_{n,m=-1}^{1} g(n, m, p)\Delta\kappa_p \sqrt{\Phi_\theta(n\Delta\kappa_p, m\Delta\kappa_p)} \exp\left[i\frac{2\pi}{N3^p}(jn + lm)\right] \quad (8.55)$$

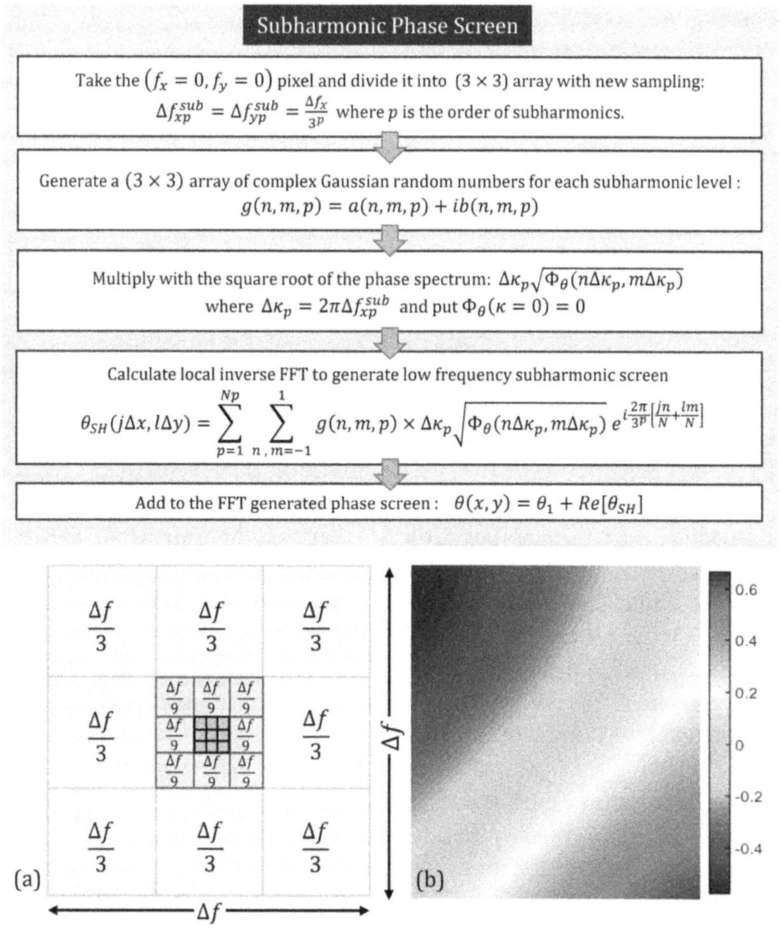

Figure 8.7. The flowchart lists the steps for generation of a sub-harmonic phase screen. The sub-harmonic grid used in calculation and a sample sub-harmonic phase screen (θ_{SH}) for $C_n^2 = 10^{-14}$ m$^{-2/3}$ is shown in (a) and (b), respectively.

where $\Delta\kappa_p = 2\pi\Delta f_{xp}^{\text{sub}}$ and $g(n, m, p)$ denotes the complex Gaussian random number for the (n, m) sub-pixel of pth sub-harmonic level. In the above summation, for each sub-harmonic level, the central sub-pixel ($m = n = 0$) representing the phase spectrum value at origin $\Phi_\theta(0, 0)$ is set equal to zero. The steps for generation of the sub-harmonic phase screen and a sample sub-harmonic screen are illustrated in figure 8.7. The final phase screen with accurate turbulence scales is then obtained by summing the sub-harmonic screen and the FFT-based phase screen:

$$\theta_{\text{Total}}(j\Delta x, l\Delta y) = \theta_{\text{FFT}}(j\Delta x, l\Delta y) + \theta_{\text{SH}}(j\Delta x, l\Delta y) \tag{8.56}$$

Let us look at the accuracy of this phase screen by comparing its expected structure function with the theoretical value. The phases θ_{FFT} and θ_{SH} are mutually independent variables with zero mean, thus from equation (8.48), one can write the 2D correlation function for θ_{Total} as [56],

$$\Gamma_{\theta_{\text{Total}}} = \Gamma_{\theta_{\text{FFT}}} + \Gamma_{\theta_{\text{SH}}} \tag{8.57}$$

The discrete 2D correlation function for the sub-harmonic phase screen is given by $\Gamma_{\theta_{\text{SH}}}$

$$\Gamma_{\theta_{\text{SH}}}(j\Delta x, l\Delta y) = \sum_{p=1}^{Np} \sum_{n,\, m\,=-1}^{1} \Phi_\theta(n\Delta\kappa_p, m\Delta\kappa_p)\ (\Delta\kappa_p)^2\ \exp\left[i\frac{2\pi}{N3^p}(jn + lm)\right] \tag{8.58}$$

The expected structure function for phase θ_{Total} can be calculated from equations (8.47), (8.57) and (8.58) using the method outlined in section (8.3.2). Figure 8.8 compares the expected structure function to the theoretical structure function D_θ^{mvK} for the cases of when sub-harmonic screens up to levels $N_p = 1$ and $N_p = 3$ were used. This figure corresponds to the same simulation parameters used for testing the FFT method-based phase screens as in figure 8.6. It can be observed that the expected structure function agrees closely with the theoretical structure function when a sufficient number of sub-harmonic levels have been included, thus pointing towards a more accurate representation of the associated turbulence scales. The case $L_o = 10$ m, $p = 3$ gives a very close match with D_θ^{mvK}, however, for $L_o = 100$ m, one needs to include more sub-harmonic levels $p > 3$. However, it has been observed that accuracy of the structure function generally saturates after a certain number of sub-harmonic levels. Adding sub-harmonic levels after $p = 15$ is usually not numerically feasible due to the overflow for double precision floating point calculations. Thus, for big (or infinite) outer scale values, the expected structure function is likely to deviate from its theoretical value at large separations. Banakh *et al* [61] studied the effectiveness of the sub-harmonic method for simulating large-scale turbulence inhomgeneities. They observed that with an increase in the number of the sub-harmonics, there was an increase in both the effective radius and relative variance of intensity fluctuations of the beam as calculated from simulation data. And these

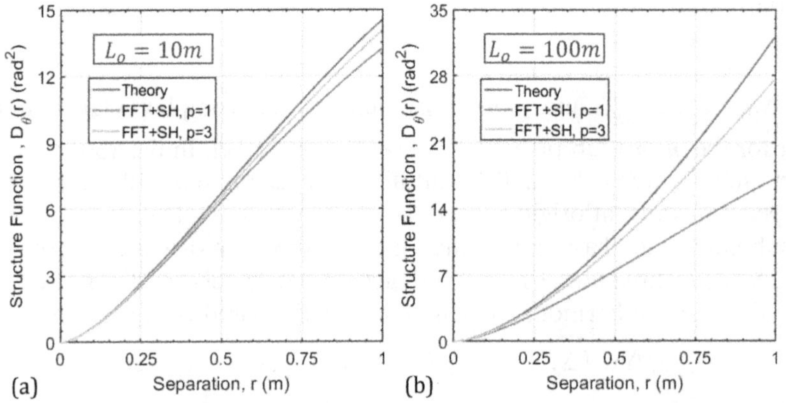

Figure 8.8. The expected structure function for the FFT method phase screens with two levels of sub-harmonic correction ($p = 1$ & $p = 3$) are compared with the theoretical structure function for outer scale values $L_o = 10\ m$ and $L_o = 100\ m$.

values had a better agreement with the theoretical and experimental data. Thus, inclusion of sub-harmonics resulted in a more accurate estimate of the beam parameters. They further observed that increase in the number of sub-harmonic levels above $N_p = 8$ was ineffective as now each higher sub-harmonic level contributed weakly compared to the previous level. Another study by Sedmak [62], provided a rough estimate for the optimum number of sub-harmonic levels that should be used. It was observed that the number of sub-harmonic levels required to obtain the optimum structure function is proportional to the ratio of the turbulence outer scale length to the screen size. Carbillet and Riccardi [63] further defined two criteria based on the integrated power over the whole range of spatial frequencies and the structure function ratio which could be used to decide the minimum number of required sub-harmonics for atmospheric turbulence phase screen simulation. Another detailed investigation for optimal sub-harmonics selection appropriate for the Kolmogorov model, the von Kármán model and the modified von Kármán model was carried out by Liu *et al* [64]. The sub-harmonic method remains a widely popular method to overcome the inadequacies in the modeling of the low frequencies in the FFT-based method by increasing the sampling density of the spatial frequencies in the vicinity of origin.

8.3.4 Some drawbacks of the FFT method

The FFT-based phase screen methods remain very popular owning to the computational efficiency of the FFT algorithm. However, there are certain disadvantages associated with this method:

1. **Under-sampling of the low spatial frequencies**:

   In the spatial domain, the phase is sampled on a fixed discrete rectangular grid. Therefore, one cannot exactly reproduce the desired phase structure function on a discrete grid as it ideally requires continuous values of the spatial frequencies κ. Besides, the central pixel corresponding to $n = m = 0$ in equation (8.45) is always put equal to zero, which leads to the under-sampling at the low spatial frequencies. The contribution of these low spatial frequencies can be accounted to a large extent by adding a sufficient number of sub-harmonics. However, in that case the FFT algorithm cannot be used for the calculation of the sub-harmonics, leading to an additional computational penalty. Also, the number of spectral components needed $(N^2 + 8N_p)$ now exceeds the number of points (N^2) in the spatial grid and could be very large for realistic phase screens.

2. **Periodic boundary conditions**:

   The phase screens are periodic with period equal to the grid size D due to the periodicity of the FFT algorithm. The periodic phase screens do not have a significant average slope [58] and so do not simulate the beam wandering correctly. This further causes difficulties in simulation of moving turbulence [65].

3. **Anisotropic phase screens**:

   For any rectangular (or square) phase screen, the loss of the low-frequency components in the diagonal direction is less than that in the x or

y direction [56]. Thus the statistical properties of the FFT-based phase screen are anisotropic in nature.

8.4 Other methods for generating random phase screens

Besides the FFT method, there are various other methods available in the literature which can be used to generate phase screens with good computational efficiency and accuracy [51, 66–75]. Goldring and Carlson [76] have reviewed five common approaches which use computational linear algebra and FFT for phase screen generation. Next, we would discuss two new recent methods in this regard.

8.4.1 Randomized spectral sampling method

Recently, Paulson, Wu and Davis presented a new modified FFT method [77] for phase screen generation based on the frequency shift property of the Fourier transform [39]. This method easily incorporates the lower spatial frequency components in the phase screen without additional computation time penalties associated with methods using sub-harmonic grids. They successfully demonstrated that for simulations of atmospheric turbulence with finite outer scales, the statistical phase structure function of the generated phase screens is in good agreement with theory. Additionally, this method can be combined with the sub-harmonic method of Lane *et al* [58] to give very accurate results across a range of spectral models of practical and theoretical interest.

Consider a phase screen simulated using the FFT-based method that is defined on a square grid as in equation (8.45). We rewrite this equation as,

$$\theta_{\mathrm{FFT}}(j\Delta x, l\Delta y) = \sum_{n,\, m=-N/2}^{N/2-1} \tilde{c}(n\Delta\kappa, m\Delta\kappa)\exp\left[i\frac{2\pi}{N}(jn + lm)\right] \tag{8.59}$$

where

$$\tilde{c}(n\Delta\kappa, m\Delta\kappa) = g(n, m)\sqrt{\Phi_\theta(n\Delta\kappa, m\Delta\kappa)}\;\;\Delta\kappa \tag{8.60}$$

Paulson *et al* [77] proposed a more meaningful use of the point closest to the κ-space origin by defining a new type of complex phase screen, θ_{PWD} such that:

$$\begin{aligned}\theta_{PWD}(j\Delta x, l\Delta y) = \sum_{n,\, m=-N/2}^{N/2-1} & \tilde{c}(n\Delta\kappa + \delta\kappa_x,\; m\Delta\kappa + \delta\kappa_y) \times \\ & \exp\left[i(j\Delta x(n\Delta\kappa + \delta\kappa_x)) + l\Delta y(m\Delta\kappa + \delta\kappa_y)\right]\end{aligned} \tag{8.61}$$

where $\delta\kappa_x$, $\delta\kappa_y$ are random variables described by a uniform distribution bounded by $\pm\Delta\kappa_x/2$ and $\pm\Delta\kappa_y/2$, respectively[2]. This offsets the lowest wave number grid point away from the origin in addition to translating the rest of the sampling grid in the frequency domain. Figure 8.9 shows the difference between the κ-space grid

[2] Note that we have $\Delta\kappa_x = 2\pi/N\Delta x$ and $\Delta\kappa_x = \Delta\kappa_y$. So, we replace $2\pi/N = \Delta\kappa\Delta x = \Delta\kappa\Delta y$ in equation (8.61).

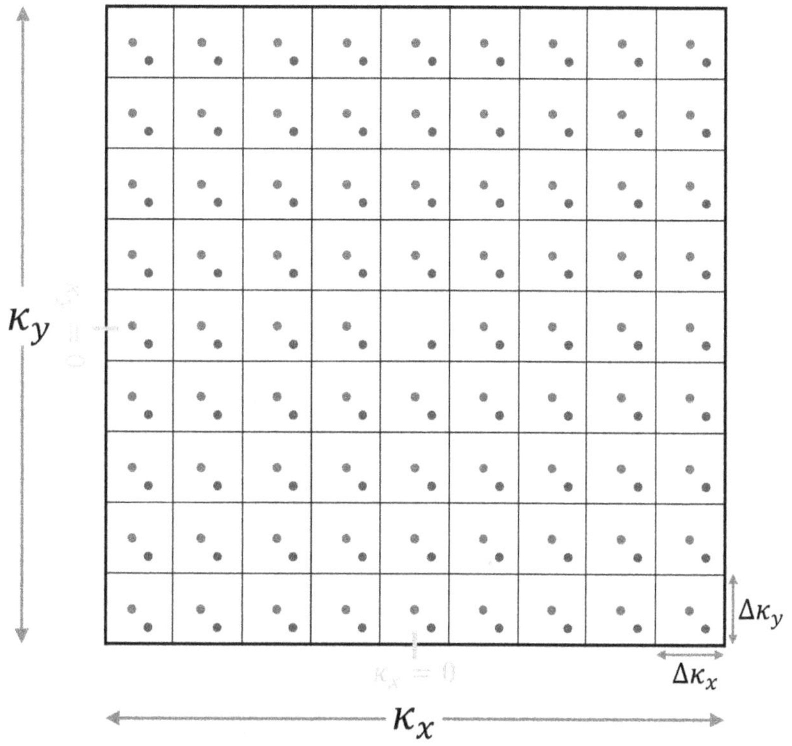

Figure 8.9. The κ-space grid sampling for the (a) traditional (blue dots) and (b) the randomized (magenta dots) spectral sampling methods is shown for one realization of the random phase screen. Note that in traditional FFT phase screens, the central pixel representing $\kappa(0, 0)$ is put to zero, whereas in the randomized spectral sampling method, a sampling point is present in the central pixel.

partitioning and sampling for the traditional FFT spectral sampling approach versus the randomized spectral sampling approach. By defining a new matrix $C(j\Delta x, l\Delta y)$ with its elements related to $\tilde{c}(n\Delta\kappa + \delta\kappa_x,\ m\Delta\kappa + \delta\kappa_y)$ by Fourier transform, one can rewrite the expression in equation (8.61) as:

$$C(j\Delta x, l\Delta y) = N^2.\ \mathcal{F}^{-1}[\tilde{c}(n\Delta\kappa + \delta\kappa_x, m\Delta\kappa + \delta\kappa_y)] \tag{8.62}$$

$$\theta_{PWD}(j\Delta x, l\Delta y) = \exp[i(j\Delta x\delta\kappa_x\ + l\Delta y\delta\kappa_y)\cdot C(j\Delta x, l\Delta y)] \tag{8.63}$$

The complex phase screen θ_{PWD} and the real-valued phase screens obtained from its real and imaginary parts, no longer exhibit periodicity and contain domain wide low spatial frequency distortions. Figure 8.10 shows the comparison between the phase screens obtained from traditional FFT method and the modified FFT method. The usual convention of setting $\tilde{c}(0, 0)$ to zero is due to the lack of a traditional spectral sampling grid point at the origin. The present method, however, provides a more meaningful use of the point closest to the κ-space origin. On further investigations, it was observed that for structure function power laws greater than the 2/3 law for Kolmogorov spectrum, the randomized spectral sampling algorithm alone was

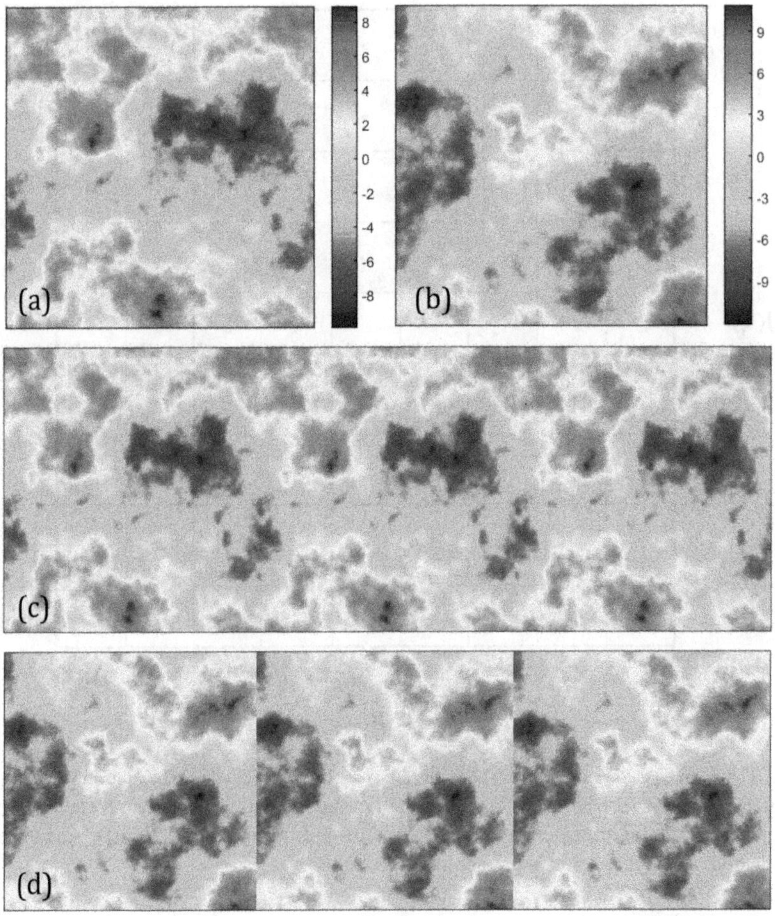

Figure 8.10. Sample phase screens are shown for (a) the traditional FFT method and (b) the modified FFT method based on randomized spectral sampling. The above phase screens are generated using a 512×512 simulation grid with von Kármán phase spectrum for $C_n^2 = 10^{-13}\,\mathrm{m}^{-2/3}$, inner and outer scale equal to $l_0 = 1$ cm, $L_0 = 3$ m, respectively, and the propagation step length $\delta z = 50$ m. To illustrate the periodic and aperiodic nature of these two types of phase screens shown in (a) and (b), we have stacked three identical copies of these phase screens adjacent to each other in (c) and (d), respectively. The aperiodic nature of the phase screens generated by using the new modified FFT method is clearly visible in (d).

actually not sufficient to ensure accurate statistics of observed simulated structure function. For this, a new hybrid algorithm utilizing both FFT-based frequency sampling randomization and sub-harmonic frequency sampling randomization was proposed [77, 78].

8.4.2 Sparse spectrum method

The sparse spectrum (SS) model was first proposed by Charnotskii for modeling sea surface elevation [79]. This technique was later introduced as an efficient way of simulating the phase front perturbations caused by atmospheric turbulence [78, 80, 81].

This model preserves the wide range of scales typically associated with atmospheric turbulence perturbations with relatively less computational efforts. The technique is based on the assumption that an individual phase samples contain only a limited number of spectral components and both the amplitude and wave vectors of these components are random. Therefore, the requirement of number of spectral components is significantly lower than the number of spatial points in the phase screen.

On similar lines as in conventional FFT-based models, let the 2D complex random phase $\theta_{SS}(\boldsymbol{r})$ be expressed as the sum of harmonics,

$$\theta_{SS}(\boldsymbol{r}) = \sum_{n=1}^{N} a_n \exp(i\boldsymbol{\kappa_n} \cdot \boldsymbol{r}) = \theta_1(\boldsymbol{r}) + i\theta_2(\boldsymbol{r}) \tag{8.64}$$

where a_n denote the random complex spectral amplitudes, $\kappa_n = (\kappa_{xn}, \kappa_{yn})$ denote the wave vectors of individual spectral components, and N represents the total number of spectral components used to represent phase [78]. In order to represent statistically homogeneous, zero average random phase, the first and second moments of the random spectral amplitudes a_n are taken as:

$$\langle a_n \rangle = 0, \ \langle a_n a_m \rangle = 0 \ \text{and} \ \langle a_n a_m^* \rangle = s_n \delta_{nm} \tag{8.65}$$

The wave vectors $\boldsymbol{\kappa_n}$ are assumed to be random vectors with probability distributions,

$$P(\boldsymbol{\kappa_n} \in (\boldsymbol{\kappa} + d\boldsymbol{\kappa})) = p_n(\boldsymbol{\kappa}) d\boldsymbol{\kappa} \tag{8.66}$$

and the ensemble averaged structure function for the phase screens of SS model is simply,

$$D_\theta^{SS}(\boldsymbol{r}) = \langle [\theta_{SS}(\boldsymbol{r}_1) - \theta_{SS}(\boldsymbol{r}_2)]^2 \rangle_{\{a_n, \boldsymbol{\kappa_n}\}} \tag{8.67}$$

$$D_\theta^{SS}(\boldsymbol{r}) = \left\langle \sum_{n=1}^{N} s_n [1 - \cos(\boldsymbol{\kappa_n} \cdot \boldsymbol{r})] \right\rangle_{\{\boldsymbol{\kappa_n}\}} \tag{8.68}$$

$$D_\theta^{SS}(\boldsymbol{r}) = \iint d^2\kappa \sum_{n=1}^{N} s_n p_n(\boldsymbol{\kappa}) [1 - \cos(\boldsymbol{\kappa_n} \cdot \boldsymbol{r})] \tag{8.69}$$

This would match the target structure function given in equation (8.41) if the following condition is satisfied:

$$\sum_{n=1}^{N} s_n p_n(\kappa) = 2\Phi_\theta(\boldsymbol{\kappa}) \tag{8.70}$$

The above equation decides the constraints on weights s_n and the probability distribution $p_n(\boldsymbol{\kappa})$. The conventional FFT phase screen model can still be obtained from the SS model in the form of equation (8.64) when $p_n(\boldsymbol{\kappa})$ are delta functions supported at the nodes of the rectangular grid [80]. For isotropic spectra $\Phi_\theta(\kappa)$, the wave vector's probability density functions $p_n(\kappa)$ can be chosen in polar coordinates:

$$p_n(\boldsymbol{\kappa})d^2\kappa = p_n(\kappa, \phi)\kappa d\kappa d\phi = p_n(\kappa)d\kappa\frac{d\phi}{2\pi} \tag{8.71}$$

The above equation assumes that the wave vector's directions are uniformly distributed on the interval $[-\pi, +\pi]$ in line with the statistical isotropy of the phase spectrum. There can be a number of different ways to satisfy equation (8.70), and each one of them will result in different Monte Carlo models for generation of the phase samples with the given structure function. Charnotskii has discussed two types of models: (a) the partitioned model (PM), and (b) the overlap model (OM), which are mathematically quite different from each other [81]. Here, we will briefly discuss the partitioned model. For the PM model, a non-overlapping partition of the desired range of wave numbers $\kappa_{\min} < \kappa < \kappa_{\max}$ into N sub-intervals is created so that

$$\kappa_{\min} < \kappa_1 < \kappa_2 < \cdots < \kappa_{N-1} < \kappa_N = \kappa_{\max} \tag{8.72}$$

and the probability distribution $p_n(\kappa)$ is supported only on the nth sub-interval $\kappa_{n-1} < \kappa < \kappa_n$. Hence,

$$s_n = 2\pi \int_{\kappa_{n-1}}^{\kappa_n} \kappa d\kappa \;\; \Phi_\theta(\kappa), \tag{8.73}$$

and,

$$p_n(\kappa) = \begin{Bmatrix} \kappa \;\; \Phi_\theta(\kappa)\left[\int_{\kappa_{n-1}}^{\kappa_n} \kappa d\kappa \;\; \Phi_\theta(\kappa)\right]^{-1}, & \kappa_{n-1} < \kappa < \kappa_n \\ 0, & \kappa < \kappa_{n-1} \;\text{ or }\; \kappa > \kappa_n \end{Bmatrix} \tag{8.74}$$

Log-uniform partition of wave numbers has been identified as the most efficient for turbulence phase screen modeling [81]. Thus, the partition wave numbers in equation (8.74) are expressed as,

$$\kappa_n = \kappa_{\min} \;\; \exp\left[\frac{n}{N}\ln\left(\frac{\kappa_{\max}}{\kappa_{\min}}\right)\right] \tag{8.75}$$

Generating practical realization of random wave numbers with probability distributions $p_n(\kappa)$ has been shown to be straightforward for pure power-law spectra [79, 81]. For more complicated spectra such as the von Kármán spectrum, such realizations can be computationally expensive. A major disadvantage of the sparse spectrum method is that it requires direct series summation of terms and one cannot make use of the the FFT algorithm to calculate the phase realizations. However, as the method makes used of moderate number of spectral components in the sparse spectrum series representation, the overall computational time for several mega-pixel-size phase screens is not too large [79]. Also, the phase samples produced using this method are unbiased and non-periodic.

8.5 Illustration of propagation of OAM states through turbulence

As an illustration of the computational methods described in this section we now show one realization of propagation of OAM states with $l = 0, 1, 2, 3$ in figure 8.11. OAM states have received much attention as communication channels for free-space communication systems [82, 83]. Rigorous turbulence modeling may also be needed for understanding degradation of entanglement when one or both the entangled photons travel through turbulence [84, 85]. As the interest in propagating individual OAM states or their superposition through realistic atmospheric conditions is growing, it is important for a number of researchers to quickly model such turbulence propagation in order to test the performance of their system designs of interest. In the illustration shown below, the inner and outer turbulence scale values of $l_o = 1$ cm and $L_o = 3$ m were used and the turbulence strength was assumed to be

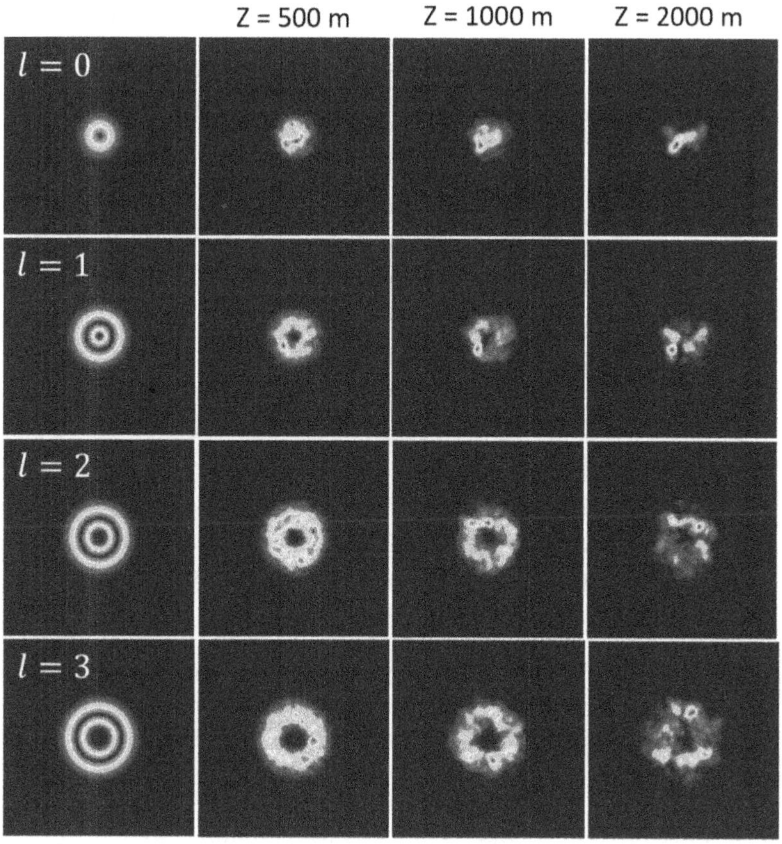

Figure 8.11. Illustration of split-step method showing one realization of propagation of OAM states $l = 0, 1, 2, 3$ through atmospheric turbulence. The beam waists parameter used for all the beams is the same and given by $W_o = 5$ cm and wavelength of the laser is $\lambda = 1.55\ \mu$m. The turbulence strength is given by $C_n^2 = 10^{-14}\ \mathrm{m}^{-2/3}$ and the von Kármán spectrum has been used. The beam intensity profiles at distances $z = 500$ m, 1000 m and 2000 m are shown. All the patterns are normalized individually for maximum dynamic range of display in individual sub-figures.

$C_n^2 = 10^{-14}\ \mathrm{m}^{-2/3}$ along with von Kármán spectrum. The sampling grid is 512 × 512 pixels with sampling interval of 1 mm. The laser wavelength is $\lambda = 1.55\ \mu\mathrm{m}$. The initial intensity profile of the beam is shown in the left column of figure 8.11, followed by intensity profiles at a propagation distance of 500 m, 1000 m and 2000 m, respectively. The split-step method used 20 random phase screens placed over a 2 km distance. The illustration shows that the intensity profile of the OAM states is fairly undisturbed till 500 m but is highly degraded over 2 km propagation distance. Detection and separation of OAM modes as required in applications such as free-space communication can therefore pose challenges over long-range turbulence paths. In the next chapter we will provide discussion on how collinear propagation of OAM states (particularly the $l = 0, 1$ states) can provide robust laser beam designs that are able to maintain their central intensity lobes on long-range turbulence propagation. The annotated software codes as described in appendix A can be implemented quickly to get an idea of beam intensity profile on propagation through turbulence. It may be noted that while beam intensity profiles are shown in figure 8.11, the phase of the propagated fields is available simultaneously. Such computed phase profiles can, for example, be utilized for estimating the nature of phase correction that may be required in an adaptive optical system.

8.6 Beam quality parameters

The amplitude and phase profile of the laser beam gets distorted on propagation through the random phase screen. A few parameters like the amount of beam spread and beam wander, scintillation index (SI), signal-to-noise ratio (SNR), bit-error-rate, average intensity, M^2 factor and the spatial coherence of the beam can be used to quantify the quality of the distorted beam. Two of these methods are described in the following discussion.

8.6.1 Scintillation index

Consider a laser beam propagating in the atmosphere and falling onto a receiver situated in any z plane. If we attempt to measure its intensity (or irradiance) at the receiver, we will find that its instantaneous intensity $I(x, y, z)$ will fluctuate in time about its average value $\langle I(x, y, z)\rangle$. These fluctuations in the beam's irradiance are known as scintillations [86]. The beam scintillations arise namely due to two reasons: (a) the beam intensity profile on passing through the random phase screen breaks up into randomly shaped speckles, and (b) the beam experiences random deflections resulting in wandering of its centroid position. Beam scintillation leads to power losses at the receiver resulting in significantly low SNR and eventually signal fading. Therefore, it is desirable to be able to predict the magnitude of intensity scintillation as an important design consideration.

The statistics concerning irradiance fluctuations and phase fluctuations are obtained from the fourth moment of the optical field. The scintillation index (or SI) is defined at any given transverse position (x, y) as:

$$\sigma_I^2(x, y, z) = \frac{\langle I(x, y, z)^2 \rangle}{\langle I(x, y, z) \rangle^2} - 1, \tag{8.76}$$

where the average intensity $\langle I(x, y, z) \rangle$ is obtained by summing over independently obtained realizations of intensity values and then dividing the result by the total number of realizations. Similarly for $\langle I(x, y, z)^2 \rangle$, the square of the intensity values of independent realizations is first added together and then averaged. The usual practice is to look at the on-axis SI i.e., when $x = y = 0$. For plane waves, the irradiance $I(x, y, z)$ at the a point detector centered at $(x = 0, y = 0)$ is enough to calculate the on-axis SI. However, in real applications, the detectors have some aperture size which gives rise to the aperture averaging effects, thus reducing the measured intensity fluctuations [87–89]. In this case, the nature of the observed intensity fluctuations depend on the size of the receiver aperture. At a given instant of time, the observable quantity then becomes,

$$S(z) = \int_A I(x, y, z) dx dy \tag{8.77}$$

where A is the surface area of the detector aperture and $I(x, y, z)$ is the instantaneous irradiance. The SI is now computed using $S(z)$,

$$\sigma_S^2 = \frac{\langle S^2 \rangle}{\langle S \rangle^2} - 1 \tag{8.78}$$

Thus, the detector aperture averages the fluctuations of the received beam over the aperture area, leading to reduced signal fluctuations compared to a point detector. The on-axis scintillation index describes the intensity fluctuations at a single on-axis point in the receiver plane. However, intensity fluctuations at one point of the beam are correlated with those at another point. This spatial structure of scintillation is studied using the co-variance function of irradiance [1]. A characteristic correlation width can be defined as either the first zero or the $1/e^2$ point of the normalized co-variance function. Aperture sizes of the order of correlation width or smaller will act like a 'point' receiver, whereas apertures greater than correlation width will cause reduction in observed scintillation due to aperture averaging. Correlation length is related to the size of the first Fresnel zone which is given by $\sqrt{L/k}$. In practice, a detector aperture of diameter less than the Fresnel length is treated as a point detector for measuring scintillation. Therefore, in simulation studies, the usual practice is to choose a detector aperture whose radius should be below $\sqrt{\lambda L/2\pi}$.

The theoretical aspects of beam scintillation have been studied over several decades, starting with Tatarskii [3, 4]. The theory of optical scintillation for plane, spherical and Gaussian beams have been established in weak turbulence conditions and the review of the same can be found in [86].

8.6.2 Signal-to-noise ratio for instantaneous beam profile

The SNR of the beam intensity is defined as the mean intensity divided by the standard deviation of intensity values within a given detector area,

$$SNR = \frac{\langle I \rangle}{\sigma_I} \quad (8.79)$$

The SNR is calculated over instantaneous intensity values which is then averaged over the total number of realizations. It is important to note here that the on-axis SI conventionally makes use of the long-time averaged intensity values while SNR uses instantaneous intensity values. In the next chapter on robust beam engineering, we will find that the instantaneous SNR is very useful in understanding the evolution of structured light beams through turbulence.

We conclude by noting that well-annotated computer simulation codes for the main numerical methods discussed in this chapter are provided in appendix A. They can be handy for beginning researchers who wish to perform simulation tests for propagation of beams with arbitrary amplitude-phase and polarization profile through turbulence.

References

[1] Andrews L C and Phillips R L 2005 *Laser Beam Propagation through Random Media* (Bellingham, WA: SPIE Press Book)

[2] Levy M 2000 *Parabolic equation Methods for Electromagnetic Wave Propagation* IEE electromagnetic wave series 45 *(Institution of Engineering and Technology)*

[3] Tatarskii V I 1961 *Wave Propagation in a Turbulent Medium* ed R A Silverman (New York: McGraw-Hill)

[4] Tatarskii V I 1971 *The effects of the turbulent atmosphere on wave propagation* (Washington, D.C., Springfield: National Oceanic and Atmospheric Administration, U.S. Department of Commerce and the National Science Foundation)

[5] Wheelon A D 2001 *Electromagnetic Scintillation* **vol 1** (Cambridge: Cambridge University Press)

[6] Fante R L 1975 Electromagnetic beam propagation in turbulent media *Proc. IEEE* **63** 1669–92 12

[7] Strohbehn J W 1978 *Laser beam propagation in the atmosphere Topics in Applied Physics* **vol 25** (Berlin: Springer)

[8] Strohbehn J W 1968 Line-of-sight wave propagation through the turbulent atmosphere *Proc. IEEE* **56** 1301–18 8

[9] Belmonte A 2000 Feasibility study for the simulation of beam propagation: consideration of coherent lidar performance *Appl. Opt.* **39** 5426–45 10

[10] Prokhorov A M, Bunkin F V, Gochelashvily K S and Shishov V I 1975 Laser irradiance propagation in turbulent media *Proc. IEEE* **63** 790–811

[11] Frehlich R G 1987 Intensity covariance of a point source in a random medium with a Kolmogorov spectrum and an inner scale of turbulence *J. Opt. Soc. Am.* A **4** 360–6 2

[12] Wang G Y and Dashen R 1993 Intensity moments for waves in random media: three-order standard asymptotic calculation *J. Opt. Soc. Am.* A **10** 1226–32 6

[13] Dashen R and Wang G Y 1993 Intensity fluctuation for waves behind a phase screen: a new asymptotic scheme *J. Opt. Soc. Am.* A **10** 1219–25 6

[14] Feizulin Z I and Kravtsov Y A 1967 Broadening of a laser beam in a turbulent medium Radiophys. *Quantum Electron.* **10** 33–5 1

[15] Lutomirski R F and Yura H T 1971 Propagation of a finite optical beam in an inhomogeneous medium *Appl. Opt.* **10** 1652–8 7

[16] Charnotskii M 2015 Extended Huygen's Fresnel principle and optical waves propagation in turbulence: discussion *J. Opt. Soc. Am.* A **32** 1357–65 7

[17] Tyson R K 2016 *Principles of Adaptive Optics* 4th edn (Boca Raton, FL: CRC Press)

[18] Roggemann M C and Welsh B M 1996 *Imaging Through Turbulence* (Boca Raton, FL: CRC Press)

[19] Marchuk G I 1980 *The Monte Carlo Methods in Atmospheric Optics*Springer series in Optical sciences *(Berlin: Springer)*

[20] Uscinski B J 1993 Multi-phase-screen analysis *Wave Propagation in Random Media (Scintillation). Proceedings of the Conference Held 3-7 August, 1992 at the University of Washington, USA* ed V Tatarskii, A Ilshimaru and V U Zavorotny (Taylor and Francis) p 346

[21] Macaskill C and Ewart T E 1984 Computer simulation of two-dimensional random wave propagation *IMA J. Appl. Math.* **33** 1–15 07

[22] Spivack M and Uscinski B J 1989 The split-step solution in random wave propagation *J. Comput. Appl. Math.* **27** 349–61

[23] Schmidt J D 2010 *Numerical Simulation of Optical Wave Propagation with Examples in MATLAB* (Bellingham, WA: SPIE Press Book)

[24] Martin J M and Flatte S M 1988 Intensity images and statistics from numerical simulation of wave propagation in 3-D random media *Appl. Opt.* **27** 2111–26

[25] Coles W A, Filice J P, Frehlich R G and Yadlowsky M 1995 Simulation of wave propagation in three-dimensional random media *Appl. Opt.* **34** 2089–101 4

[26] Johansson E M and Gavel D T 1994 Simulation of stellar speckle imaging *Amplitude and Intensity Spatial Interferometry II* **vol 2200** ed J B Breckinridge (Bellingham, WA: International Society for Optics and Photonics, SPIE) pp 372–83

[27] Knepp D L 1983 Multiple phase-screen calculation of the temporal behavior of stochastic waves *Proc. IEEE* **71** 722–37 6

[28] Martin J M and Flatté S M 1990 Simulation of point-source scintillation through three-dimensional random media *J. Opt. Soc. Am.* A **7** 838–47

[29] Voelz D G and Xiao X 2009 Metric for optimizing spatially partially coherent beams for propagation through turbulence *Opt. Eng.* **48** 1–7

[30] Dockery G D 1988 Modeling electromagnetic wave propagation in the troposphere using the parabolic equation IEEE Trans *Antennas Propag* **36** 1464–70

[31] Kuttler J R and Dockery G D 1991 Theoretical description of the parabolic approximation/Fourier split-step method of representing electromagnetic propagation in the troposphere *Radio Sci.* **26** 381–93

[32] Fleck J A, Morris J R and Feit M D 1976 Time-dependent propagation of high energy laser beams through the atmosphere *Appl. Phys.* **10** 129–60

[33] Weideman J A C and Herbst B M 1986 Split-step methods for the solution of the nonlinear Schrödinger equation *SIAM J. Numer. Anal.* **23** 485–507

[34] Leontovich M A and Fock V A 1946 Solution of the problem of propagation of electromagnetic waves along the Earth's surface by method of parabolic equations *J. Phys. USSR* **10** 13–23

[35] Hardin R H and Tappert F D 1973 Applications of the split-step Fourier method to the numerical solution of nonlinear and variable coefficient wave equations *SIAM Rev.* **15** 423

[36] Flatte S M and Tappert F D 1975 Calculation of the effect of internal waves on oceanic sound transmission *J. Acoust. Soc. Am.* **58** 1151–9
[37] DiNapoli F R and Deavenport R L 1977 *Numerical Methods of Underwater Acoustic Propagation* (Berlin: Springer)
[38] Tappert F D 1974 Parabolic equation method in underwater acoustics *J. Acoust. Soc. Am.* **55** S34
[39] Goodman J 2004 *Introduction to Fourier Optics* 3rd edn (Viva Books)
[40] Voelz D G and Roggemann M C 2009 Digital simulation of scalar optical diffraction: revisiting chirp function sampling criteria and consequences *Appl. Opt.* **48** 6132–42 11
[41] Liu J-P 2012 Controlling the aliasing by zero-padding in the digital calculation of the scalar diffraction *J. Opt. Soc. Am.* A **29** 1956–64 9
[42] Mas D, Garcia J, Ferreira C, Bernardo L M and Marinho F 1999 Fast algorithms for free-space diffraction patterns calculation *Opt. Commun.* **164** 233–45
[43] Lochab P P, Senthilkumaran P and Khare K 2019 Propapation of converging polarization singular beams through atmospheric turbulence *Appl. Opt.* **58** 6335–45
[44] Matsushima K and Shimobaba T 2009 Band-limited angular spectrum method for numerical simulation of free-space propagation in far and near fields *Opt. Express* **17** 19662–73
[45] Yura H T 1979 Signal-to-noise ratio of heterodyne lidar systems in the presence of atmospheric turbulence *J. Opt.* **26** 627–44
[46] Xiao X and Voelz D 2009 On-axis probability density function and fade behavior of partially coherent beams propagating through turbulence *Appl. Opt.* **48** 167–75 1
[47] Bramley E N, Appleton and Victor E 1954 The diffraction of waves by an irregular refracting medium *Proc. R. Soc.* **225** 515–8
[48] Ratcliffe J A 1956 Some aspects of diffraction theory and their application to the ionosphere *Rep. Prog. Phys.* **19** 188–267 1
[49] Booker H G, Ferguson J A and Vats H O 1985 Comparison between the extended-medium and the phase-screen scintillation theories *J. Atmos. Terr. Phys.* **47** 381–99
[50] Uscinski B J 1985 Analytical solution of the fourth-moment equation and interpretation as a set of phase screens *J. Opt. Soc. Am.* A **2** 2077–91
[51] Welsh B M 1997 Fourier-series-based atmospheric phase screen generator for simulating anisoplanatic geometries and temporal evolution *Propagation and Imaging through the Atmosphere* **vol 3125** ed L R Bissonnette and C Dainty (Bellingham, WA: International Society for Optics and Photonics, SPIE) pp 327–38
[52] Roddier N A 1990 Atmospheric wavefront simulation using Zernike polynomials *Opt. Eng.* **29** 1174–80
[53] Mitra S K 2001 *Digital Signal Processing: A Computer-Based Approach* 2nd edn (New York: McGraw-Hill)
[54] Herman B J and Strugala L A 1990 Method for inclusion of low-frequency contributions in numerical representation of atmospheric turbulence *Propagation of High-Energy Laser Beams Through the Earth's Atmosphere* **vol 1221** ed P B Ulrich and L E Wilson (Bellingham, WA: International Society for Optics and Photonics, SPIE) 183–92 pp
[55] Noll R J 1976 Zernike polynomials and atmospheric turbulence *J. Opt. Soc. Am.* **66** 207–11 3
[56] Xiang J 2012 Accurate compensation of the low-frequency components for the FFT-based turbulent phase screen *Opt. Express* **20** 681–7 1

[57] Vorontsov A M, Paramonov P V, Valley M T and Vorontsov M A 2008 Generation of infinitely long phase screens for modeling of optical wave propagation in atmospheric turbulence *Waves Random Complex Media* **18** 91–108
[58] Lane R G, Glindemann A and Dainty J C 1992 Simulation of a Kolmogorov phase screen *Waves Random Media* **2** 209–24
[59] Frehlich R 2000 Simulation of laser propagation in a turbulent atmosphere *Appl. Opt.* **39** 393–7
[60] Roggemann M C 2014 *Simulating non-Kolomogorov phase screens with finite inner and outer scales Imaging and Applied Optics 2014* (New York: Optical Society of America) p PM4E.1
[61] Banakh V A, Krekov G M, Mironov V L, Khmelevtsov S S and Tsvik R S 1974 *J. Opt. Soc. Am.* **64** 516–8 4
[62] Sedmak G 2004 Implementation of fast-Fourier-transform-based simulations of extra-large atmospheric phase and scintillation screens *Appl. Opt.* **43** 4527–38 8
[63] Carbillet M and Riccardi A 2010 Numerical modeling of atmospherically perturbed phase screens: new solutions for classical fast Fourier transform and Zernike methods *Appl. Opt.* **49** G47–52 11
[64] Liu T, Zhang J, Lei Y, Zhang R and Sun J 2019 Optimal subharmonics selection for atmosphere turbulence phase screen simulation using the subharmonic method *J. Mod. Opt.* **66** 986–91
[65] Glindemann A, Lane R G and Dainty J C 1993 Simulation of time-evolving speckle patterns using Kolmogorov statistics *J. Mod. Opt.* **40** 2381–8
[66] Harding C M, Johnston R A and Lane R G 1999 Fast simulation of a Kolmogorov phase screen *Appl. Opt.* **38** 2161–70 4
[67] Johnston R A and Lane R G 2000 Modeling scintillation from an aperiodic Kolmogorov phase screen *Appl. Opt.* **39** 4761–9 9
[68] Roggemann M C, Welsh B M, Montera D and Rhoadarmer T A 1995 Method for simulating atmospheric turbulence phase effects for multiple time slices and anisoplanatic conditions *Appl. Opt.* **34** 4037–51 7
[69] Beghi A, Cenedese A and Masiero A 2008 Stochastic realization approach to the efficient simulation of phase screens *J. Opt. Soc. Am.* A **25** 515–25 2
[70] Sriram V and Kearney D 2007 An ultra fast Kolmogorov phase screen generator suitable for parallel implementation *Opt. Express* **15** 13709–14 10
[71] Dios F, Recolons J, Rodríguez A and Batet O 2008 Temporal analysis of laser beam propagation in the atmosphere using computer-generated long phase screens *Opt. Express* **16** 2206–20 2
[72] Kouznetsov D, Voitsekhovich V V and Ortega-Martinez R 1997 Simulations of turbulence-induced phase and log-amplitude distortions *Appl. Opt.* **36** 464–9 1
[73] Eckert R J and Goda M E 2006 Polar phase screens: a comparative analysis with other methods of random phase screen generation *Atmospheric Optical Modeling, Measurement, and Simulation II* **vol 6303** ed S M Hammel and A Kohnle (Bellingham, WA: International Society for Optics and Photonics, SPIE) pp 1–14
[74] Jakobsson H 1996 Simulations of time series of atmospherically distorted wave fronts *Appl. Opt.* **35** 1561–5 3
[75] Assémat F, Wilson R W and Gendron E 2006 Method for simulating infinitely long and non stationary phase screens with optimized memory storage *Opt. Express* **14** 988–99 2

[76] Goldring T and Carlson L 1989 Analysis and implementation of non-Kolmogorov phase screens appropriate to structured environments *Nonlinear Optical Beam Manipulation and High Energy Beam Propagation Through the Atmosphere* **vol 1060** ed R A Fisher and L E Wilson (Bellingham, WA: International Society for Optics and Photonics, SPIE) pp 244–64
[77] Paulson D A, Wu C and Davis C C 2019 Randomized spectral sampling for efficient simulation of laser propagation through optical turbulence *J. Opt. Soc. Am.* B **36** 3249–62 11
[78] Charnotskii M 2020 Comparison of four techniques for turbulent phase screens simulation *J. Opt. Soc. Am.* A **37** 738–47 5
[79] Charnotskii M 2011 Sparse spectrum model of the sea surface *30th Int. Conf. on Ocean, Offshore and Arctic Engineering* **vol 6** (Ocean Engineering (ASME))
[80] Charnotskii M 2013 Sparse spectrum model for a turbulent phase *J. Opt. Soc. Am.* A **30** 479–88
[81] Charnotskii M 2013 Statistics of the sparse spectrum turbulent phase *J. Opt. Soc. Am.* A **30** 2455–65 12
[82] Malik M, O'Sullivan M, Rodenburg B, Mirhosseini M, Leach J, Lavery M P J, Padgett M J and Boyd R W 2012 Influence of atmospheric turbulence on optical communications using orbital angular momentum for encoding *Opt. Express* **20** 13195–200
[83] Krenn M, Fickler R, Fink M, Handsteiner J, Malik M, Scheidl T, Ursin R and Zeilinger A 2014 Communication with spatially modulated light through turbulent air across Vienna *New J. Phys.* **16** 113028
[84] Roux F S 2011 Infinitesimal-propagation equation for decoherence of an orbital-angular-momentum-entangled biphoton state in atmospheric turbulence *Phys. Rev.* A **83** 053822
[85] Zhang Y *et al* 2016 Experimentally observed decay of high-dimensional entanglement through turbulence *Phys. Rev.* A **94** 032310
[86] Andrews L C, Phillips R L and Young C Y 2001 *Laser Beam Scintillation with Applications* (Bellingham, WA: SPIE)
[87] Fried D L 1967 Aperture averaging of scintillation *J. Opt. Soc. Am.* **57** 169–75
[88] Andrews L C 1992 Aperture-averaging factor for optical scintillations of plane and spherical waves in the atmosphere *J. Opt. Soc. Am.* A **9** 597–600
[89] Wang S J, Baykal Y and Plonus M A 1983 Receiver-aperture averaging effects for the intensity fluctuation of a beam wave in the turbulent atmosphere *J. Opt. Soc. Am.* **73** 831–7 6

IOP Publishing

Orbital Angular Momentum States of Light (Second Edition)
Propagation through atmospheric turbulence
Kedar Khare, Priyanka Lochab and Paramasivam Senthilkumaran

Chapter 9

Robust laser beam engineering using complementary diffraction

This chapter introduces concepts that utilize polarization and orbital angular momentum (OAM) (or phase) degrees of freedom that lead to beam designs that show robust intensity profiles on propagation through turbulence. Several applications in defense and communication require the overall beam intensity profile to be smooth at the receiver end. While adaptive optics has been considered a traditional solution to this problem, an alternative of engineering the beam profile offers some new avenues for future applications as discussed here.

9.1 Beam engineering using polarization and OAM

This chapter describes a robust beam engineering principle that has resulted from the recent work of the authors of this book. The intensity profile of optical beams on encountering spatially varying refractive index gets distorted beyond the nominal beam spreading due to diffraction effects. Refractive index variations of the order of 10^{-5}–10^{-6} are sufficient to make the beam profile speckled. This poses a serious problem for multiple applications like free-space optical (FSO) communication, underwater communication, laser-guided defense systems, light detection and ranging (LiDAR), etc which depend on laser beam propagation through a random medium like the atmosphere, or ocean. The performance of these systems is greatly limited because the random time-varying fluctuations in the beam intensity compromises the amount of energy that can be delivered by the beam. This beam degradation essentially leads to signal fading and a broken optical link. This long-standing problem of sending information or energy through a random medium by optical beams has attracted a lot of attention from the scientific community. Various types of beam designs dealing with the beam properties like beam profile, polarization, coherence etc have been proposed and tested [1–8]. However, it is not easy to provide specific guidelines for generating robust beams for propagation through a

doi:10.1088/978-0-7503-5959-7ch9

random medium due to the stochastic nature of the problem. In this chapter, we take a step in this direction and provide a novel idea for designing robust vector beams using the speckle diversity principle. The basic aim is to evolve an optimal beam engineering approach using the interplay between polarization and OAM states of light that is inherently best suited for sending light beams through a turbulent medium.

In this chapter we will first discuss a peculiar property associated with diffraction patterns observed when a general aperture is illuminated using the $l = 0$ and $l = 1$ OAM states and use it further to develop a robust beam engineering principle that explicitly uses the polarization singular beams explained in chapter 6. Our simulations and experiments suggest that this principle may be used to design beams that maintain their intensity profile even after propagation through random phase fluctuations as encountered in the atmosphere. For simplicity the atmosphere will be modeled first using a single phase screen. Next we will validate the robust beam engineering principle for long-range turbulence propagation using the split-step method.

9.2 Complementary diffraction due to (0,1) OAM states

In chapter 2 we developed the notion of the spiral phase quadrature transform as a two-dimensional analogue of the Hilbert transform. It was shown that two functions $g_1(x, y)$ and $g_2(x, y)$ connected by the spiral phase filter $\exp(i\phi)$ (ϕ is the polar angle in two-dimensional Fourier space) approximately had sine–cosine-like behavior in the sense of Mandel's theorem. The simplest way to realize the above result in an optics laboratory is to illuminate an arbitrary amplitude-phase aperture with $l = 0$ and $l = 1$ OAM states that distinctly differ in their phase profile which is given by $\exp(il\phi)$. The result that the spiral phase filter is an analogue of Hilbert transform implies that the Fraunhofer diffraction patterns observed due to these two illuminations for a given aperture must show sine–cosine-like diversity. We illustrate this effect first with diffraction patterns associated with simple apertures. Figure 9.1 shows the Fraunhofer diffraction intensity patterns corresponding to circular, square and rectangular (slit-like) apertures when illuminated by the $l = 0$ and $l = 1$ OAM states. In particular, we observe the maxima and minima in the two diffraction intensity patterns due to the two OAM states getting exchanged. It should be observed that this complementarity is local just like in a sine–cosine pair. As we will see later in this section, the complementary diffraction is simply a property of the spiral phase quadrature transform and should, therefore, also be observed for general random phase apertures as well. One case of interest to multiple applications is the propagation of optical beams through atmospheric turbulence. A random medium like atmosphere at a given instant can be approximately described by a suitable phase screen in the lens aperture. Therefore, we next simulate the diffraction of these two OAM states through a random phase screen. In order to generate the random phase screen, we first take a random phase function in the lens aperture with phase distributed uniformly in $[0, 2\pi]$. This random phase function is then convolved with an averaging Gaussian filter to induce correlation over phase

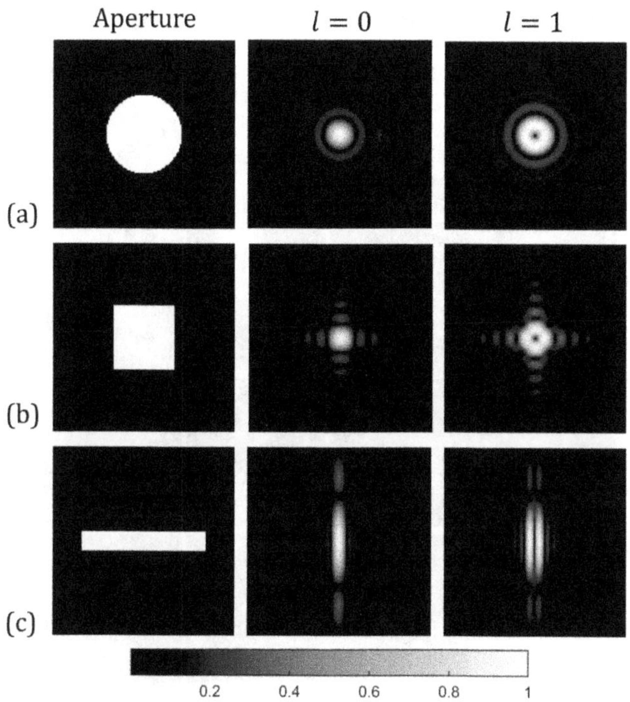

Figure 9.1. Diffraction patterns corresponding to the $l = 0$ and $l = 1$ OAM states are shown for (a) circular (b) square and (c) rectangular apertures, respectively.

fluctuation length scale. Later on in the chapter, we will model the atmosphere with random phase screens generated using the FFT method, as explained in chapter 8. For now, the present method of generating random phase screen is sufficient for illustration of complementarity. Figure 9.2 shows the diffraction patterns corresponding to $l = 0$ and $l = 1$ OAM states when random phase was introduced in the lens aperture plane. Individually, the diffraction spot has degraded to speckled appearance due to random phase aperture. The complementary nature of these diffraction patterns can, however, be easily observed for the case of random phase screen apertures as well. In order to quantify the complementarity, we calculate the correlation coefficient (ρ_{I_1,I_2}) between the intensities I_1 and I_2 corresponding to the OAM states $l = 0$ and $l = 1$. The correlation coefficient is defined as:

$$\rho_{I_1,I_2} = \frac{\langle[(I_1 - \mu_1)(I_2 - \mu_2)]\rangle}{\sigma_{I_1}\sigma_{I_2}} \tag{9.1}$$

where $\langle\ \rangle$ denote the ensemble average and μ and σ are the mean and the standard deviation, respectively. The correlation coefficient is a measure of the strength of linear association between two variables. It can take any value in the range $[-1, +1]$ with $+1$ (or -1) indicating perfect positive (or negative) correlation. The positive correlation means that any increase (or decrease) in the value of one variable is accompanied by increase (or decrease) in the value of the second variable.

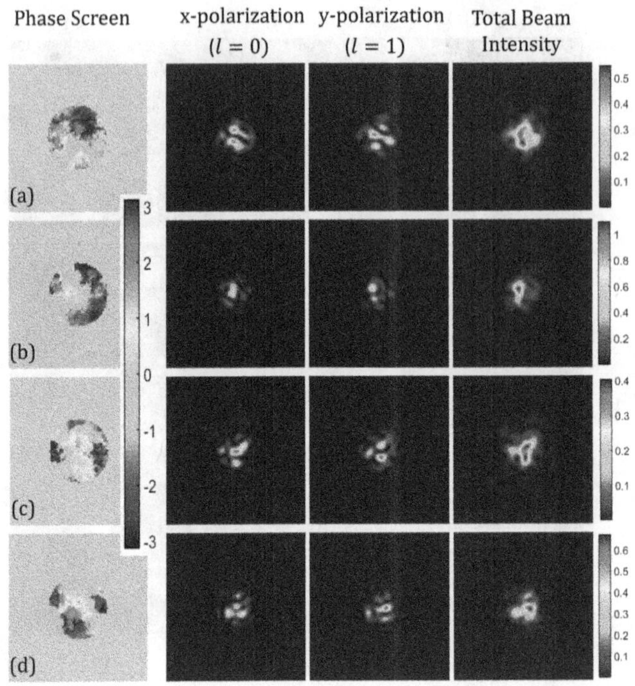

Figure 9.2. The simulated intensities corresponding to the $l = 0$, 1 illumination are shown for four different random phase screens which are placed in the lens aperture.

A negative correlation on the other hand means that one variable increases as the second variable decreases and vice versa. A value of zero would correspond to no correlation between the two variables. In the present case, the diffraction patterns of the $l = 0$ and $l = 1$ OAM states would be complementary if the corresponding correlation coefficient has a sufficiently large negative value. Based on 50 realizations of the random phase screen, the correlation coefficient for the intensity images for $l = 0$ and $l = 1$ OAM states is observed to be equal to $\rho_{I_1,I_2} = -0.95 \pm 0.04$. This value of ρ_{I_1,I_2} has been evaluated over a region equal to the free-space diffraction-limited spot size of the $l = 0$ OAM mode. Thus, the two diffraction intensity patterns are in fact highly negatively correlated. This high negative correlation suggests that if the diffraction patterns corresponding to $l = 0$ and $l = 1$ OAM states are added together, the resulting total intensity profile should be more uniform than the individual diffraction pattern intensities. The last column in figure 9.2 shows this combined (or total) beam intensity profile. It can be observed that the combined intensity profile has a well-defined central lobe even when different random phase screens are used. Therefore, when the $l = 0$ and $l = 1$ OAM states are propagated collinearly through a given random phase screen, the complementary diffraction intensity patterns when added together can yield an improved intensity profile. In practice, the random phase screens may represent different realizations of a fluctuating atmosphere that are time-varying. The complementary diffraction effect will still remain valid thereby maintaining beam quality through time-varying random phase fluctuations.

It is important to note that incoherent addition of two diffraction intensity patterns is required for this robust beam engineering approach. The incoherent addition can be achieved, for example, by engineering the beam such that the two OAM states are embedded in two orthogonal polarizations. Another possible mechanism is to combine two independent laser beams made to operate in the $l = 0$ and $l = 1$ spatial modes. In the present work, we will take the polarization approach for convenience. This approach will also allow us naturally to describe the engineered beams in terms of the polarization singularities formalism. Choosing the $l = 0$ OAM state to be in x polarization and $l = 1$ OAM state in the y polarization, the engineered vector beam propagating in z direction could be described in the $z = 0$ plane as,

$$\vec{E}(x, y) = \frac{1}{\sqrt{2}}[\hat{\boldsymbol{x}} + e^{i\theta}\hat{\boldsymbol{y}}]\psi(x, y) \tag{9.2}$$

where the function $\psi(x, y)$ represents the normalized beam profile function, for example, the Gaussian or LG(0,0) mode or any other native beam profile. A time dependence of form $\exp(-i\omega t)$ is assumed. Here, $\hat{x}$ and $\hat{y}$ represent unit vectors in the x and y directions, respectively, and θ is the polar angle. To experimentally realize this beam, one may use the arrangement involving a spatial light modulator, as shown in figure 9.3. A 45-degree linearly polarized laser beam in $l = 0$ OAM state is made incident on a reflective phase spatial light modulator (SLM). The liquid

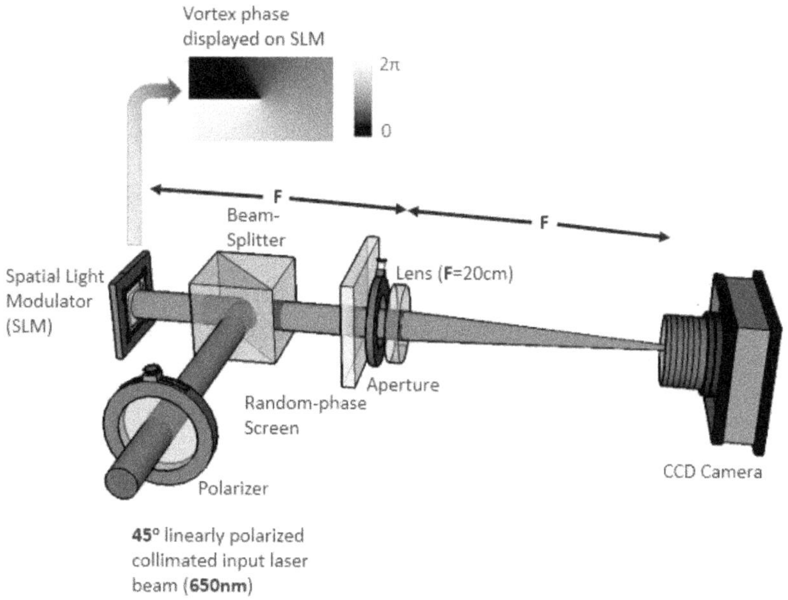

Figure 9.3. Experimental set-up for generation of vector beam obtained by combining $l = 0, 1$ OAM states in orthogonal polarizations. Reprinted with permission from [9] © 2017 Optical Society of America.

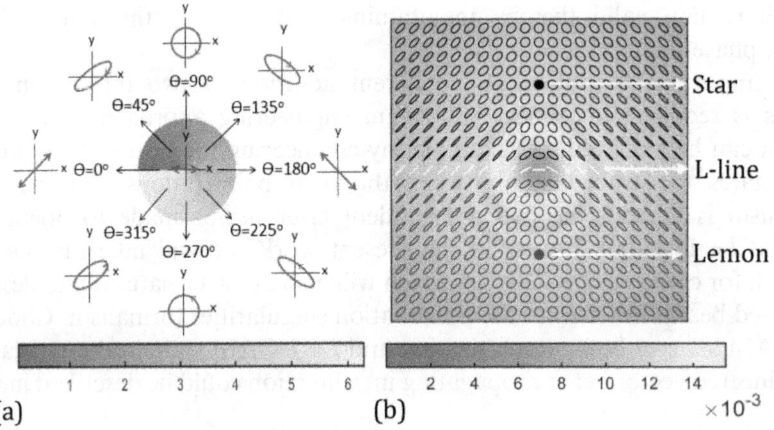

Figure 9.4. The state of polarization of the engineered vector beam is shown (a) at the SLM plane and (b) after free-space propagation from the SLM plane. The color map in (a) shows the phase difference between the E_x and E_y orthogonal polarization components of the vector beam. The polarization states in (b) are plotted on top of the simulated engineered beam intensity. The black and blue colors in (b) represent the opposite handedness of the polarization ellipses. The polarization singularity structures such as lemon, star and L-line are also marked. Adapted with permission from [9] © 2017 The Optical Society of America.

crystal SLMs are usually sensitive to one polarization and act like a plane mirror for the orthogonal polarization. The vortex phase θ is displayed on the SLM. On reflection from the SLM, both the polarization components get reflected collinearly with one of the polarizations in $l = 1$ OAM state to generate the beam described in equation (9.2). The polarization structure of the reflected beam at the SLM plane and after propagation from SLM plane is shown in figure 9.4 From the discussion in chapter 6, it is clear that the above vector beam contains C-point polarization singularity structures (lemon, star separated by L-line) embedded in it after it has propagated from the SLM plane. The polarization map has been generated after numerically propagating the scalar components in equation (9.2). If the individual scalar components are passed through the random phase screen shown in figure 9.3, the diffraction pattern may be observed on an array sensor placed in the back focal plane of the Fourier transforming lens. Let us denote the transmission function of the phase screen in the lens aperture by $t(x, y)$, then the total intensity of the engineered beam as observed on the charge-coupled device (CCD) sensor in the back focal plane of the lens may be described (apart from constants) as:

$$I_{\text{Total}}(u, v) = I_1 + I_2 = |\mathcal{F}\{t(x, y)E_x(x, y)\}|^2 + \left|\mathcal{F}\{t(x, y)E_y(x, y)\}\right|^2 \qquad (9.3)$$

where E_x and E_y denote the x and y components of the field in the random phase screen plane. It is worth noting that the electric fields corresponding to the x and y polarizations in the camera plane essentially have a two-dimensional Hilbert transform relationship due to the OAM diversity in E_x and E_y. The coordinates in the camera plane are denoted by (u, v). The operation $\mathcal{F}$ denotes the two-dimensional Fourier transform operation corresponding to the Fraunhofer

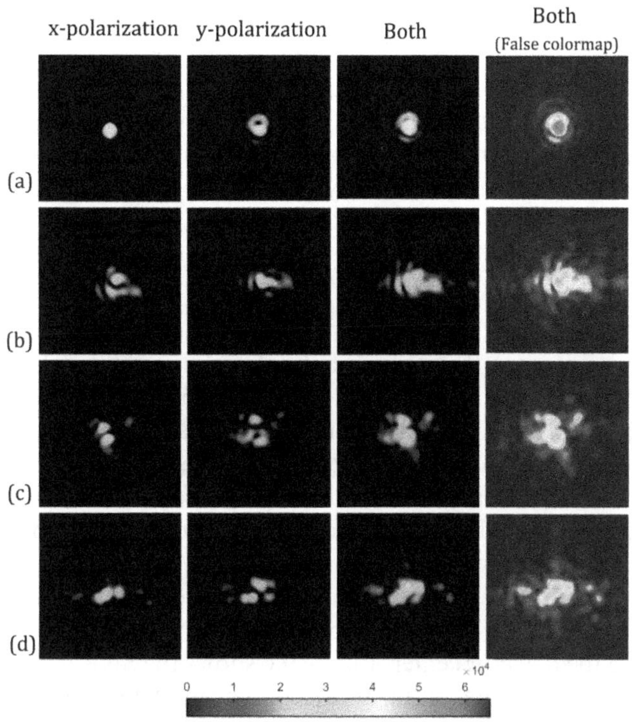

Figure 9.5. The experimentally recorded intensities for vector beam (given in equation (9.2)) along with its constituent orthogonal polarizations are shown for three different realizations of the random phase screen in (b–d). Here, different regions of a crumpled polythene sheet have been used as the random phase screens. The first row (a) shows the recorded intensities when the beam passes through open aperture. The last column shows the total intensity of the vector beam in false colors. Adapted with permission from [9] © 2017 The Optical Society of America.

diffraction. In writing the total intensity as a sum of the two polarizations, we have neglected the cross-talk between the two orthogonal polarization components as it is usually very negligible in the case of thin random phase screens [10] and becomes important only when dealing with thick scattering media. The intensity patterns I_1 and I_2 for individual polarizations may be observed on the sensor by introducing a polarizer in the beam path. The experimental intensity profiles corresponding to the simulated intensity patterns shown in figure 9.2 are shown in figure 9.5. In this experiment, crumpled transparent polythene sheets were used as random phase screens [9]. The smooth total intensity patterns for various random phase screens shown in figure 9.5 suggest that examining the diversity in I_1 and I_2 may be an interesting method for robust beam design as we will explore in the next section.

9.3 Beam quality assessment using instantaneous signal-to-noise ratio

We have discussed the concept of speckle diversity for robust beam design for propagation through random phase perturbations. In this section, we introduce a beam quality parameter—the instantaneous signal-to-noise ratio (SNR)—which provides quantitative assessment of the improvement in beam intensity profile. As we will discuss later, the on-axis scintillation index (SI) has commonly been used as a measure for beam quality, typically for Gaussian beams. The SI, however, does not reflect the nature of improvement in transverse beam profile. We define instantaneous beam SNR as the mean intensity $\langle I \rangle$ divided by the standard deviation of intensity σ_I values within a given detector area:

$$SNR(r_d) = \frac{\langle I \rangle}{\sigma_I}. \tag{9.4}$$

The parameter r_d denotes the radius of the detector area that has been used for computing the SNR. The SNR is calculated for individual realizations of the random phase screens which may then be averaged over the total number of realizations. For calculation of SNR, we first find the location of the peak intensity of the engineered vector beam. Then, a circular area centered at this peak location is selected from the beam and the SNR is calculated for the enclosed intensity values. These circular regions have been centered on the peak position of the engineered beam as this is naturally the best position of the detector in any real practical system. Typically fast kHz rate tip–tilt corrections in the beam may be made to bring the peak intensity to the detector center. Figure 9.6 shows the SNR curves for individual polarizations as well as total beam intensity for simulated (figure 9.2) as well as experimental (figure 9.5) intensity profiles. The parameter r on the x-axis of SNR plots represents the radius of the free-space diffraction-limited spot for the lens-camera Fourier transform arrangement in figure 9.3. The error bars in these curves represent standard deviations in the SNR values corresponding to the 50 realizations of the random phase screens used in simulations and experiments, respectively. It is observed that the SNR values for the engineered beam are consistently greater than those for the individual polarizations components in both simulation and experiments. At a detector radius of half the diffraction-limited spot, the beam SNR is observed to be higher by an order of magnitude or more, which may offer a significant advantage in terms of maintaining a smooth beam profile over the detector.

9.4 Speckle diversity

The recent literature has hinted at the advantage of using inhomogeneously polarized beams for propagation through turbulence [9, 11–17]. As discussed in chapter 6, inhomogeneously polarized light can be considered to be a vector superposition of distinct amplitude-phase wave functions in the two polarizations. The two orthogonally polarized states of an inhomogeneously polarized beam do not interfere and as a result, the overall intensity profile is equal to the sum of its two polarization components. If these two polarization states evolve into sufficiently diverse speckles on propagation through a random medium, then their intensity sum will have a smoother profile compared to the individual polarization states. Keeping this diversity idea in mind, we explore the propagation of vector beams in various

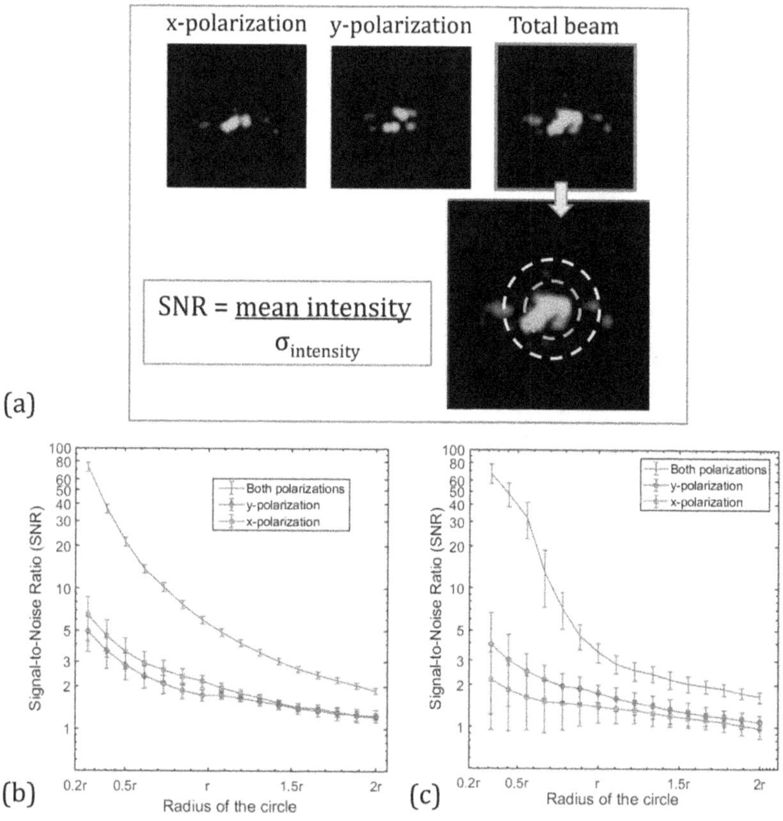

Figure 9.6. The methodology for calculating the SNR of the engineered beam is illustrated in (a). First, the point of maximum intensity of the vector beam is identified, and a circular area with a variable radius is centered at this point. The SNR is then calculated using the intensity pixels within this circular region. The variation of SNR with respect to the radius of the circle is presented in (b) for simulated intensity images and in (c) for experimentally obtained intensity images. Adapted with permission from [9] © 2017 Optical Society of America.

states of polarization through a random phase screen and observe their far-field diffraction patterns. We will specifically study the C-point and V-point polarization singular beams as they form an important class of inhomogeneously polarized beams.

Consider a vector beam propagating in the $+z$ direction where its electric field may be described in the paraxial form as,

$$\boldsymbol{E}(x, y, z) = [E_1(x, y)\hat{e}_1 + E_2(x, y)\hat{e}_2]\exp(ikz). \tag{9.5}$$

The orthogonal unit vectors $\hat{e}_1$ and $\hat{e}_2$ may denote Cartesian vectors $(\hat{x} - \hat{y})$ denoting linear polarization or the right circularly or left circularly polarized (RCP–LCP) states $(\hat{e}_R - \hat{e}_L)$. Here, a time dependence of the form $\exp(-i\omega t)$ and circular polarization basis $(\hat{e}_1 = \hat{e}_R, \hat{e}_2 = \hat{e}_L)$ has been assumed. The polarization distribution of the vector beam is decided by the functional form of the amplitude and phase of the two orthogonal polarization components $E_1(x, y)$ and $E_2(x, y)$. When these two components have different amplitude and phase structure, the resulting vector beam has spatially inhomogeneous polarization. Our aim is to understand if we can design $E_1(x, y)$ and $E_2(x, y)$ such that the total intensity pattern $I_{\text{Total}}(u, v) = I_1(u, v) + I_2(u, v)$ as in equation (9.3) can be made resistant to random phase perturbations. In this context, the diversity in the speckle pattern intensities $I_1(u, v)$ and $I_2(u, v)$ is an important factor to be taken into consideration. Clearly, if $I_1(u, v)$ and $I_2(u, v)$ are highly correlated, their addition will not lead to much gain in terms of reducing the intensity fluctuations. The traditional literature on speckle has considered reduction in speckle contrast on addition of uncorrelated speckles [18]. However, if the two speckle patterns $I_1(u, v)$ and $I_2(u, v)$ are negatively correlated, the resulting total intensity pattern would have less contrast and more uniformity. This is so because in negatively correlated speckle patterns, the location of intensity maxima for one polarization will coincide with the location of intensity minima for the other. Therefore, the addition of such complementary speckles would reduce the fluctuations in the overall intensity. Now, we study four different types of vector beams with different amplitude and phase structures for $E_1(x, y)$ and $E_2(x, y)$. These beams have been chosen such that the speckles corresponding to the $E_1(x, y)$ and $E_2(x, y)$ components span a range of correlation coefficients from +1 to −1. This is achieved by constructing vector beams with increasingly different $E_1(x, y)$ and $E_2(x, y)$ profiles in terms of their amplitude, phase and OAM. These vector beams are then propagated through a random phase screen and the correlation coefficient (see equation (9.1)) between the intensities $I_1(u, v)$ and $I_2(u, v)$ is calculated over the diffraction-limited spot size. The simulated far-field intensities of the vector beams and their orthogonal polarization components ($\hat{e}_R - \hat{e}_L$) are illustrated in figure 9.7 for a given realization of the random phase screen. The first column in the figure provides the polarization distribution of the vector beam. The mean correlation coefficient is mentioned in the last column, which has been calculated over a set of 50 independent realizations of the random phase screen. We next describe these four cases one by one.

1. **Scalar beam**: The first case shown in figure 9.7(a) corresponds to a homogeneously polarized beam with equal energy in the $\hat{e}_R$ and $\hat{e}_L$ components. Both $E_1(x, y)$ and $E_2(x, y)$ components are taken in the $l = 0$ OAM state. In this case, the speckles produced by the two orthogonal polarizations on passing through a given random phase screen are identical with perfect correlation of $\rho_{I_1,I_2} = +1$. Therefore, when the two intensities I_1 and I_2 are added, there is no advantage in terms of beam quality.
2. **Amplitude diversity**: Next, we consider the case of an inhomogeneously polarized beam obtained by using different amplitude profiles for $E_1(x, y)$ and $E_2(x, y)$ components, as shown in figure 9.7(b). The $\hat{e}_R$ polarization contains a $l = 0$ OAM mode while the $\hat{e}_L$ polarization contains a dark Hollow Gaussian Beam (HGB) [4, 19]. The HGB has a donut intensity profile similar to the $l = 1$ OAM mode but it does not carry any OAM. The HGB field at $z = 0$ plane can be expressed as:

$$E_{\text{HGB}} = A_o \exp(-r^2/W_o^2)(r^2/W_o^2)\exp(ikz) \tag{9.6}$$

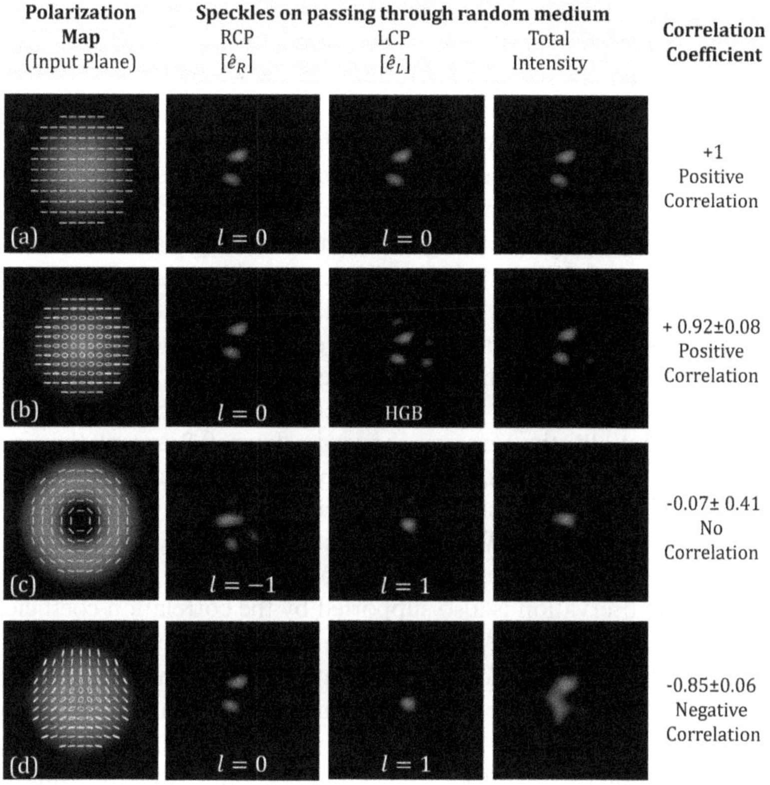

Figure 9.7. Simulation results of beams with different polarization structures and their corresponding far-field intensity profiles after passing through a random phase screen are illustrated. The intensities I_1 and I_2 for RCP and LCP components of the beams are also displayed. The correlation coefficient ρ_{I_1,I_2} is averaged over 50 realizations of the random phase screen. Note that 'HGB' stands for Hollow Gaussian Beam. Reprinted with permission from [12], copyright (2018) by the American Physical Society.

where $r = \sqrt{x^2 + y^2}$ is the radial coordinate in the transverse plane, W_o is the beam waist and A_o is the normalization constant of the $l = 0$ state. Therefore, the vector beam can be written as,

$$\boldsymbol{E}(x, y) = A_o \exp(-r^2/W_o^2)\,[\hat{e}_R + (r^2/W_o^2)\hat{e}_L]\exp(ikz) \tag{9.7}$$

For this beam, we find that the intensities I_1 and I_2 are positively correlated with a correlation coefficient value of $\rho_{I_1,I_2} = 0.92 \pm 0.08$. Therefore, the vector beam intensity profile would show similar fluctuations as observed for the individually polarized components. This means that any inhomogeneously polarized beam formed as a result of only amplitude diversity in the two polarizations components is not necessarily useful in maintaining a robust intensity through turbulence.

3. **Phase diversity**: For the third case, we add phase diversity in the two polarization states in the form of different OAM states $l = -1$ and $l = 1$,

given by $\exp(-i\theta)$ and $\exp(i\theta)$, respectively. The two polarization states now contain phase singularity of same strength but opposite handedness. Embedding the $l = -1$ and $l = 1$ states in the $\hat{e}_R$ and $\hat{e}_L$ polarization states leads to the V-point polarization singularity in the form of the azimuthal polarization. The intensity profiles for this beam and its constituent orthogonal polarizations is shown is figure 9.7(c). It is observed that the correlation coefficient between the intensity profiles I_1 and I_2 is $\rho_{I_1,I_2} = -0.07 \pm 0.41$ which indicates uncorrelated speckle patterns on an average. One can see from the figure 9.7(c) that the speckle contrast in this case reduces for the vector beam as compared to the individual $\hat{e}_R - \hat{e}_L$ states.

4. **OAM diversity**: For the last case, we have used $l = 0$ and $l = 1$ OAM states in the $\hat{e}_R$ and $\hat{e}_L$ polarizations. This choice of OAM states in the orthogonal polarization generates a star-type C-point polarization singularity. This beam is essentially the same as shown in figure 9.2 except that instead of $(\hat{x} - \hat{y})$ polarization basis, the current beam has circular polarization basis $(\hat{e}_R - \hat{e}_L)$. Figure 9.7(d) shows the intensity profiles of this beam and its constituent components. We observe that the intensity profile of the vector beam has significantly improved compared to the $l = 0$ and $l = 1$ OAM states. This observation is also supported by the correlation coefficient value $\rho_{I_1,I_2} = -0.85 \pm 0.06$ which shows high negative correlation between the I_1 and I_2 profiles. The high negative correlation can also be visibly seen in the I_1 and I_2 intensity patterns where the maxima of one polarization complements the minima of the other polarization. This complementarity property is unique to C-point structures. As a result, the complementary evolution of the two polarization structures can help maintain the total intensity of the beam, even when propagating through a time-varying random medium.

The polarization singularity structures described above may be generated in the laboratory using the methods described in chapter 6 by vector superposition of scalar waveforms in orthogonal polarizations. The above analysis shows that not all inhomogeneously polarized beams are equally robust against phase perturbations. Only specific designs of the vector beams whose orthogonal polarizations produce either uncorrelated or negatively correlated speckle patterns offer an improvement in the total intensity profile of the vector beam. Therefore, an intelligent choice of $E_1(x, y)$ and $E_2(x, y)$ should be made in the vector beam design. In this regard, the speckle diversity in the intensity profiles of the orthogonal polarization is an important feature which can be used as a general guideline for designing robust beams.

The experimentally recorded intensities [12] for the V-point and C-point singular beams after they have passed through the random phase screen are shown in figure 9.8. Panels (a) and (b) in the figure correspond to V-point singular beams, namely the azimuthal and radial polarizations while panels (c) and (d) show the lemon and star-type C-point polarization singular beams, respectively. The decomposition of these beams in circular basis is also illustrated. The figure shows the

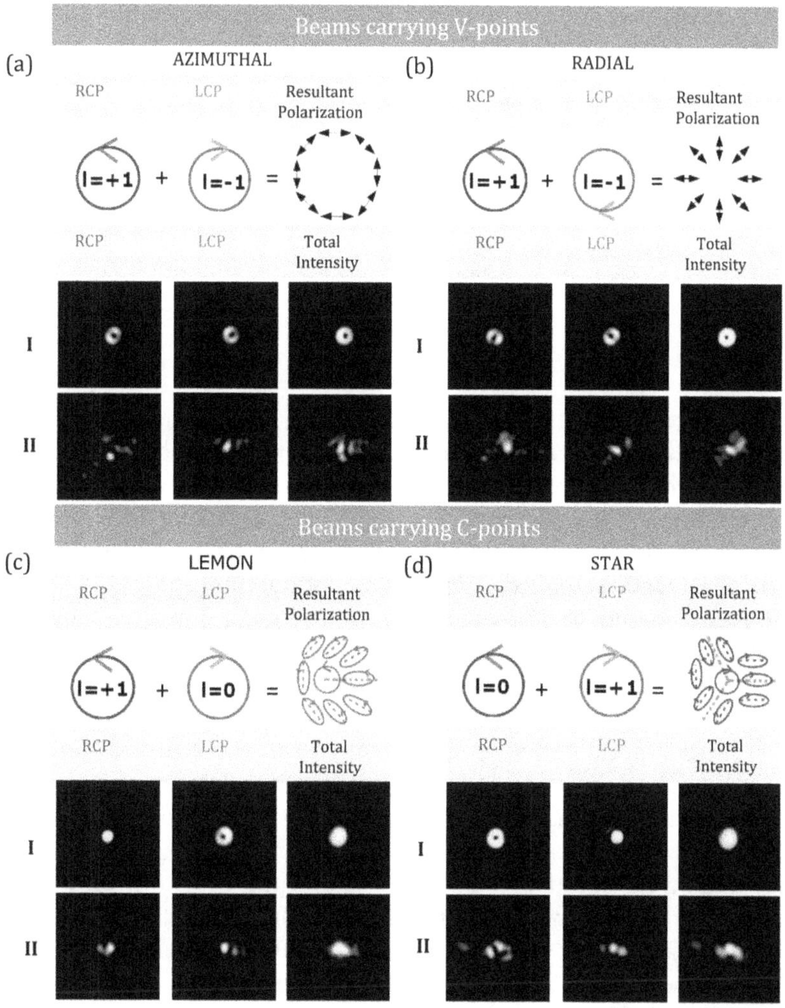

Figure 9.8. The experimentally obtained intensities for beams carrying V-point and C-point polarization singularities are shown in (a–b) and (c–d), respectively. Panels (a) and (b) depict the azimuthal and radial V-point singular beams, while panels (c) and (d) illustrate the lemon and star-type C-point polarization singular beams. The decomposition of these beams into RCP and LCP components is also provided. Row I displays the Fraunhofer diffraction intensities for the case without a phase screen, while row II shows the intensities after the beam has passed through a random phase screen. Reprinted with permission from [12], copyright (2018) by the American Physical Society.

intensity I_{Total} of the vector beams and the intensities I_1 and I_2 of their two constituent polarization components. In each of the panels (a–d), row **I** shows the intensities for the beams as they pass through an open aperture without any random phase screen while row **II** shows the intensity records when a random phase screen was present in the aperture plane. It can be easily seen through visual inspection that the beam quality of the vector beams is superior when it contained a C-point polarization singular structure as compared to the cases when it had a V-point

Table 9.1. Correlation coefficient for the I_1 and I_2 intensity patterns on CCD sensor for five different types of polarization singular beams.

Type of Singularity	Singularity Structure	Correlation Coefficient
C-point	Lemon	-0.81 ± 0.06
C-point	Star	-0.76 ± 0.07
C-point	Lemon-Star dipole	-0.82 ± 0.06
V-point	Radial	0.11 ± 0.41
V-point	Azimuthal	0.19 ± 0.35

polarization singular structure. The correlation coefficient calculated between the experimentally recorded I_1 and I_2 intensities for the beams is shown in table 9.1. The standard deviation in the correlation coefficient is calculated over 50 experimental realizations of random phase screen. The quality of the V-point and C-point singular beams after propagation through random medium can be quantified using the beam SNR. The SNR curves for experimental and simulation data [12] are shown in figure 9.9. Panels (a) and (b) show the SNR curves for the simulated intensities of the V-point and the C-point polarization singular beams. These plots show the variation of the SNR values as a function of the radius of circular region used for calculation of SNR. The parameter r on the x-axis of the plots corresponds to the radius of the free-space diffraction-limited spot size. Once again, the error bars in these plots represent the standard deviation of the SNR values calculated over 50 realizations of the random phase screen. The SNR curves for the experimentally obtained intensities of the V-point and C-point polarization singular beams are shown in panels (c) and (d), respectively. Panel (e) shows the SNR values for a C-point polarization singular beam, whose polarization distribution is in the form of a lemon-star C-point dipole. The last panel (f) contains a summary of the experimentally obtained SNR results for various V-point and C-point beams for convenience. The SNR values for the scalar $l = 0$ and $l = 1$ OAM states are also shown. One main difference in this plot is that the SNR calculation for the $l = 0$ and $l = 1$ OAM beams has been carried out at their individual intensity peak position. Note that there is a slight difference in the SNR values for the $\hat{e}_R$ and $\hat{e}_L$ polarized components in the V-point and C-point polarization singular beams in figures 9.9 (a)–(d). The peak intensity in the V-point polarization singular beam typically occurs near the intensity peak of either one of its two constituent polarizations. This is understandable as the individual polarizations of the V-point singular beam containing $l = -1$ and $l = +1$ OAM states produce uncorrelated speckle patterns, as observed from table 9.1. Therefore, during the calculation of SNR values, the center of the circular region used for calculating SNR, is closer to the peaks of the I_1 or I_2 profiles, thus yielding a slightly different SNR value for the RCP–LCP components. However, for the C-point beam, the intensity profiles of I_1 and I_2 are negatively correlated, meaning that the peak intensity of the C-point polarization singular

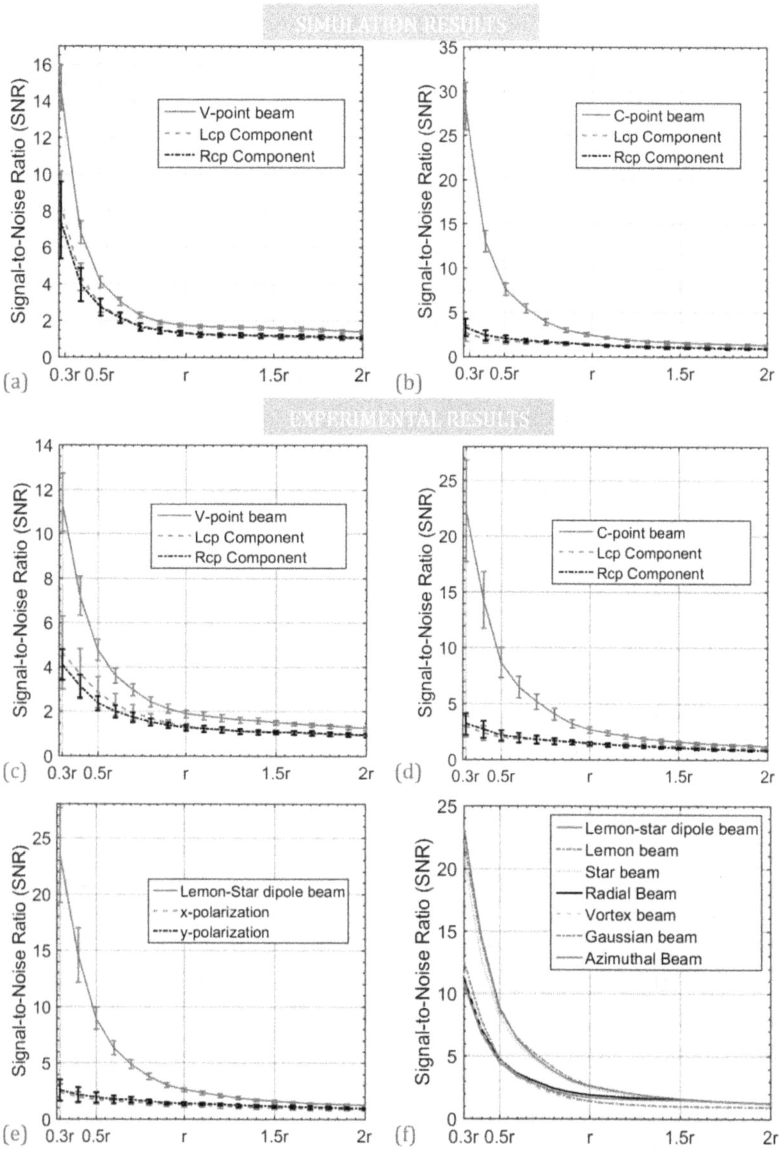

Figure 9.9. Simulated and experimentally observed SNR curves for the total beam and its RCP–LCP components are shown for V-point and C-point beams in panels (a) and (c), and panels (b) and (d), respectively. Panel (e) shows SNR curves for a beam containing C-point dipole structure. Panel (f) shows a summary SNR comparison for C-point carrying beams, V-point carrying beams, C-point dipole beam and homogeneously polarized Gaussian and a charge +1 vortex beam. Reprinted with permission from [12], copyright (2018) by the American Physical Society.

beam typically does not coincide with the peak intensities of either I_1 and I_2. Consequently, the center of the circular region used for calculating SNR is not positioned near the maxima of I_1 and I_2, which slightly reduces the measured SNR values overall.

We make some important observations from the SNR plots. First of all, the SNR for the I_{Total} is always more than that for the individual polarization intensities I_1 and I_2 for both V-point and C-point polarization singular beams. The C-point polarization singular beam shows approximately two-fold or more SNR gain over the V-point polarization singular beam when the SNR is calculated over a circle of size $0.5r$. The superior beam quality of C-point singular beams is observed regardless of the polarization basis used. The comparison of various scalar and vector beams in panel (f) highlights the importance of choosing suitable $E_1(x, y)$ and $E_2(x, y)$ fields for engineering robust vector beams. Finally, the complementary speckles of $l = 0$ and $l = 1$ OAM states provide a convenient way to enhance speckle diversity in the I_1 and I_2 profiles. This increased diversity helps ensure that the overall beam intensity experiences reduced fluctuations when passing through a turbulent random medium, such as the atmosphere.

9.5 Long-range propagation of converging polarization singularities through atmospheric turbulence

In all the illustrations in this chapter so far, the random phase fluctuations were modeled using a single random phase screen. In practice, when propagation through long-range turbulence ($\sim$a few kms) is of interest, a more appropriate model is to use the split-step method for computing field propagation, as explained in chapter 8. In particular, a number of random phase screens with known power spectrum are generated and placed along the beam path interspersed with segments of free-space propagation. In long-range propagation, it is common practice to introduce a concave curvature at the input side in order to focus the beam on the target. While the split-step method is well known, our main goal is to understand if our idea of complementary diffraction for the $l = 0$, 1 OAM states still holds when the long-range propagation is modeled with this more rigorous approach. The material provided here is largely based on the work of the present authors [20].

A number of theoretical and simulation studies of long-range propagation in atmospheric turbulence have been undertaken for Gaussian beams [21, 22]. They have been investigated in different turbulence strengths, beam parameters and focusing conditions and also as a potential ground station for satellite uplink [23–26]. Various other beam amplitude types like Bessel [27], Laguerre–Gaussian, Annular, cosh and cos Gaussian [2] have also been investigated. Other types of Gaussian beams like the dark Hollow Gaussian [4] and flat-top Gaussian [3] have also been studied. The phase singularities of optical beams are known to be resistant against phase and amplitude perturbations [28]. The topological charge of vortex beams has been shown to be conserved on propagation through the atmosphere [29] and has been useful as an information carrier in free-space optical communication. Different variants of vortex beams like the flat-topped vortex beams and different turbulence strengths have also been studied [30–32]. Propagation of OAM beams have also been explored in the context of maintaining entanglement in quantum communication [33–36]. It has been observed that the scintillation for partially polarized and partially coherent beam is less than that of a fully coherent beam [5–8]. It has been theoretically shown by Schulz

[37] that beams which have minimum scintillation are in general partially coherent. Many optimization schemes have been developed to incorporate these beams in free-space optical communication [38]. The use of vector beams with inhomogeneous polarization has been proposed [11, 12, 39]. In this respect, vector vortex beams with radial and azimuthal polarization have been studied [13–15, 40].

Polarization singular beams have been described in detail in chapter 6. The lowest order C-point and V-point beams can be written in the form:

$$\boldsymbol{E}_C(\boldsymbol{r}, z) = \frac{1}{\sqrt{2}}[re^{i\theta}\hat{\boldsymbol{e}}_R + \hat{\boldsymbol{e}}_L]\psi(\boldsymbol{r}, z) \tag{9.8}$$

$$\boldsymbol{E}_V(\boldsymbol{r}, z) = \frac{1}{\sqrt{2}}[re^{i\theta}\hat{\boldsymbol{e}}_R + re^{-i\theta}\hat{\boldsymbol{e}}_L]\psi(\boldsymbol{r}, z) \tag{9.9}$$

Here $\psi(\boldsymbol{r}, z)$ is the host beam profile given by the Gaussian or the LG(0,0) mode, and can be expressed as,

$$\psi(\boldsymbol{r}; z = 0) = A\exp\left(-\frac{r^2}{W_o^2} - \frac{i\pi r^2}{\lambda F_o}\right) \tag{9.10}$$

where A is the complex amplitude and λ denotes the wavelength. The beam waist of the Gaussian at the $z = 0$ plane is denoted by W_o and $F_o(>0)$ represents the radius of curvature of the wavefront. The beam waist at any other distance $z > 0$ can be computed using Fresnel propagation and is given by [41]:

$$W(z) = W_o\sqrt{\left(1 - \frac{z}{F_o}\right)^2 + \left(\frac{z\lambda}{\pi W_o^2}\right)^2} \tag{9.11}$$

By differentiating equation (9.11) with respect to z, the location z_f at which the beam would nominally focus in free space can be obtained as:

$$z_f = F_o\left[1 + \left(\frac{2F_o}{kW_o^2}\right)^2\right]^{-1} \tag{9.12}$$

It can be observed from equation (9.12) that $z_f < F_o$, which means that the effective focal length is shorter than the radius of the curvature of the beam [21, 42]. For beams having W_o of the order of tens of cms and F_o value in kms, this change in the focusing length is not negligible. For example, for a beam with $W_o = 7$ cm, $\lambda = 1.55\ \mu$m and $F_o = 2000$ m, $z_f = 1922$ m. In order to focus the beam at a particular z distance, equation (9.12) can be inverted and solved for F_o. This yields the expression for F_o as,

$$F_o = \frac{k^2W_o^4}{8z_f}\left[1 - \sqrt{1 - \frac{16z_f^2}{k^2W_o^2}}\right] \tag{9.13}$$

Table 9.2. Simulation parameters for propagation of polarization singular beams through atmospheric turbulence

Parameter Name	Symbol	Value
Propagation distance	z	2000 m
Radius of curvature	F_o	2019 m
Inter-screen distance	δz	100 m
Number of screens	M	20
Grid points	N	512
Grid spacing	δx	0.001 m
Beam Waist	W_o	0.1 m
Wavelength	λ	1.55 μm
Inner scale	l_o	0.01 m
Outer scale	L_o	3 m

The result in equation (9.13) is true in free space, however, the beams behave differently in turbulence. An interesting observation that has been reported is that the focus spot moves slightly towards the transmitter on atmospheric propagation [23]. In the current study, this focus shift has not been considered and all beam parameters like SNR, SI, etc are calculated at the final propagation plane situated at distance $z = 2$ km. The geometry of propagation is as shown in figure 8.2 except that now the input beam has additional concave curvature. In the following illustrations we have used the von Kármán spectrum for representing the refractive index fluctuations with inner scale $l_o = 0.01$ m and outer scales $L_o = 3$ m. The refractive index structure constant is taken as $C_n^2 = 1 \times 10^{-14}$ m$^{-2/3}$ and $C_n^2 = 1 \times 10^{-13}$ m$^{-2/3}$ for representing moderate and high level of turbulence while the propagation distance z is kept equal to 2 km. The various parameters used for simulation are listed in table 9.2.

9.5.1 Evolution of intensity and polarization structure of the beams

The total intensity for the polarization singular beam on propagation can be obtained by adding individual intensities of $\hat{e}_R$ and $\hat{e}_L$ polarizations. This is approximately true for media like atmosphere where the cross-talk between the two orthogonal polarization components may be neglected in low angle forward propagation. The total intensity profile of C-point and V-point polarization singular beams at different propagation distances is shown in figures 9.10(a) and 9.11(a) for $C_n^2 = 10^{-14}$ m$^{-2/3}$. These figures illustrate the intensity evolution of the two beams for a specific realization of atmospheric turbulence. At $z = 2$ km, it is evident that the instantaneous intensity profile of the C-point beam is more uniform compared to that of the V-point beam. The polarization structures of these beams along with the corresponding Stokes phases are also illustrated in figures 9.10(b and c) and 9.11(b and c), respectively. The Stokes phase (as discussed in chapter 6) is characteristically different for V-point and C-point polarization singular structures and therefore,

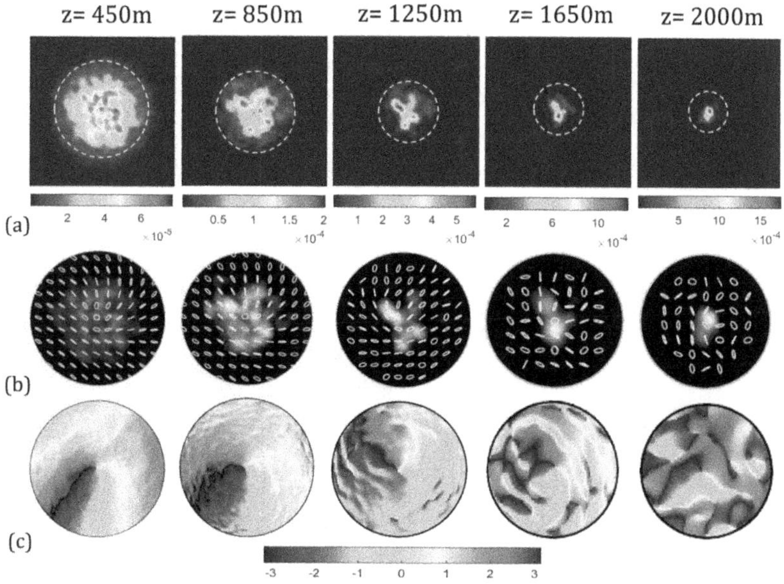

Figure 9.10. The figure illustrates the evolution of the C-point polarization singular beam as it propagates through moderate turbulence, characterized by $C_n^2 = 1 \times 10^{-14}\ \mathrm{m}^{-2/3}$. Panel (a) displays the intensity profile of the C-point beam at various propagation distances (z), with each window size set to 320 × 320 pixels. Panels (b) and (c) show the polarization pattern and the corresponding Stokes phase of the beam within the region enclosed by the dotted white circle. Note that the polarization pattern is overlaid on the beam's intensity profile. Adapted with permission from [20] © 2019 Optical Society of America.

offers a great visual aid for identification of the polarization singularities in the beam. For example, as shown in figures 9.10(c) and 9.11(c), we observe that for a C-point beam, the Stokes phase has only 1 phase cut and the total phase variation is from 0 to 2π while the Stokes phase for the V-point beam has 2 phase cuts each going from 0 to 2π. The C-point beam shown here has a Lemon type polarization singular structure in the initial plane while the V-point beam has an azimuthal polarization structure.

A few observations can be made from the polarization and Stokes phase distributions of these beams on long-range propagation through turbulence. The polarization distribution of the C-point beam shows a clockwise rotation as z increases as seen in figure 9.10(b). This has been reported earlier for the lowest order C-point beam when propagated in free space [43] and in linear and non-linear media [44] where the polarization distribution is seen to undergo a rigid rotation of $\pi/2$ as beam propagates from waist plane to the far-field zone. In figure 9.10(c), the Stokes phase also shows this clockwise rotation of the polarization pattern. The Stokes phase distribution of the V-point beam, as seen in figure 9.11(c), shows that near $z = 1250$ m, the two oppositely charged vortex components in the V-point beam probably split into two separate vortices. At this point the polarization distribution also get scrambled. Thus, due to the atmospheric turbulence, the two orthogonal polarization components of the V-point beam, $l = 0$ and $l = 1$ OAM

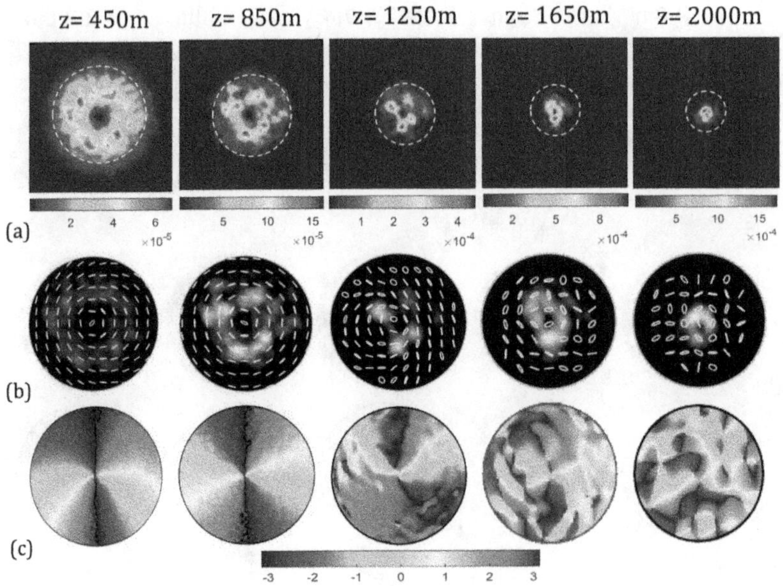

Figure 9.11. The figure illustrates the evolution of the V-point polarization singular beam as it propagates through moderate turbulence, characterized by $C_n{}^2 = 1 \times 10^{-14}\ \mathrm{m}^{-2/3}$. Panel (a) displays the intensity profile of the V-point beam at various propagation distances (z), with each window size set to 320 × 320 pixels. Panels (b) and (c) show the polarization pattern and the corresponding Stokes phase of the beam within the region enclosed by the dotted white circle. Note that the polarization pattern is overlaid on the beam's intensity profile. Adapted with permission from [20] © 2019 Optical Society of America.

states evolve differently. This may result in a lateral shift in the position of the vortices of the $l = 1$ and $l = -1$ OAM states, thus resulting in two spatially separated cuts in the Stokes phase of the V-point beam. In general, the polarization distribution of both V-point and C-point beams becomes increasingly scrambled with greater propagation distance. However, the Stokes phase structure of the C-point beam is better preserved compared to that of the V-point beam.

The intensity evolution and polarization distribution of C-point and V-point beams at different z values is shown in figure 9.12 for the case of high turbulence, characterized by $C_n^2 = 1 \times 10^{-13}\ \mathrm{m}^{-2/3}$. A few observations can be drawn from the illustrations. In this case, the beams experience more spreading due to higher phase fluctuations. Observations of the beam spot across multiple realizations reveal that a portion of the C-point beam maintains a uniform intensity area within the spot, whereas no corresponding uniform intensity area is observed in the V-point beam for the same turbulence realizations. The intensity profiles of the beams at $z = 1650$ m appear more focused than the intensity profiles at $z = 2000$ m. It is known that in turbulence, the beam focuses before its geometrical focus. However, this shifting of focal point towards the transmitter seems to increase with increasing C_n^2 value. Note for comparison, for the beam sizes in figures 9.10 and 9.11, where we see that for $C_n^2 = 1 \times 10^{-14}\ \mathrm{m}^{-2/3}$, the beams focus point is near the geometrical focus situated at 2000 m, whereas this is not true when C_n^2 is taken to be $1 \times 10^{-13}\ \mathrm{m}^{-2/3}$.

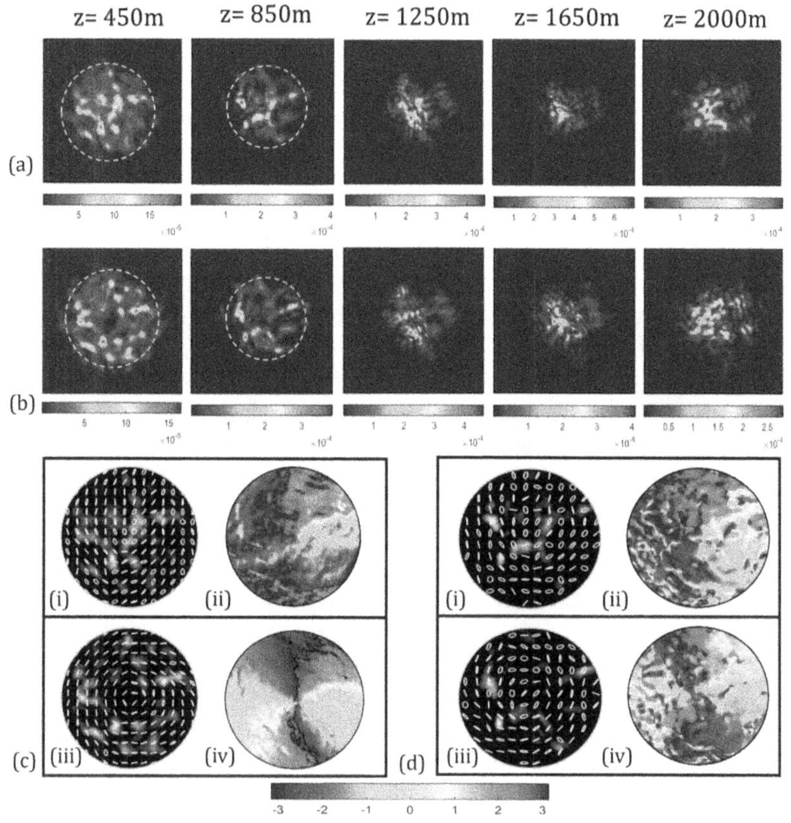

Figure 9.12. The figure illustrates the evolution of the intensity and polarization structure of polarization singular beams as they propagate through high turbulence levels, characterized by $C_n^2 = 1 \times 10^{-13}\ \mathrm{m}^{-2/3}$. Panels (a) and (b) show the intensity profiles of the C-point and V-point polarization singular beams at various z values, with each window size being 320 × 320 pixels. The polarization pattern and the corresponding Stokes phase of the beams at propagation distances $z = 450$ m and $z = 850$ m for the region marked by a dotted white circle are shown in (c) and (d), respectively. In both (c) and (d), (i–ii) correspond to C-point beam while (iii–iv) are for V-point beam. Note that the polarization pattern is shown superimposed on the respective beam's intensity profiles.

Therefore, in higher turbulence, the radius of curvature of the beam might need to be further adjusted. Figures 9.12(c) and (d) show the polarization distribution and the corresponding Stokes phases for the beams at $z = 450$ m and $z = 850$ m, respectively. It is clearly visible that the polarization structure of the beam is severely distorted by $z = 850$ m.

9.5.2 Quantitative assessment of beam quality

Next, we calculate the SNR and on-axis SI in order to quantitatively measure the beam quality. Earlier studies on focused non-singular beams have shown that they have less scintillation near the focus point, however, these beams are more susceptible to beam wander [25, 26, 45–48]. Therefore, in order to have any actual reduction in scintillation on the receiver, it is necessary that these beams should be actively corrected for beam wander. Therefore, we report SNR and on-axis SI for both cases of tracked and untracked beams. The beam position is usually centered on the detector by using a tip/tilt correction to counteract the effects of beam wander. Fast scanning mirrors with kHz tip–tilt correction rate may be required for correcting the beam wander.

We use an on-axis circular detector for calculation of SNR for the untracked beams. In the case of tracked C-point and V-point beams, the circular detector is always centered on their instantaneous peak intensity positions. For the same positions of the detector, SNR values for the intensity profiles of the individual orthogonal polarization states are also calculated. We have varied the radius of the detector from 25% to 150% of the free-space diffraction-limited spot-radius of the Gaussian beam, as given by equation (9.11). The variation of SNR of the C-point and V-point beams with increasing detector radius is shown in figures 9.13(a) and (b) for the untracked beams and in figures 9.13(c) and (d) for the tracked beams for $C_n^2 = 1 \times 10^{-14}$ m$^{-2/3}$. The parameter 'r' here signifies the free-space diffraction-

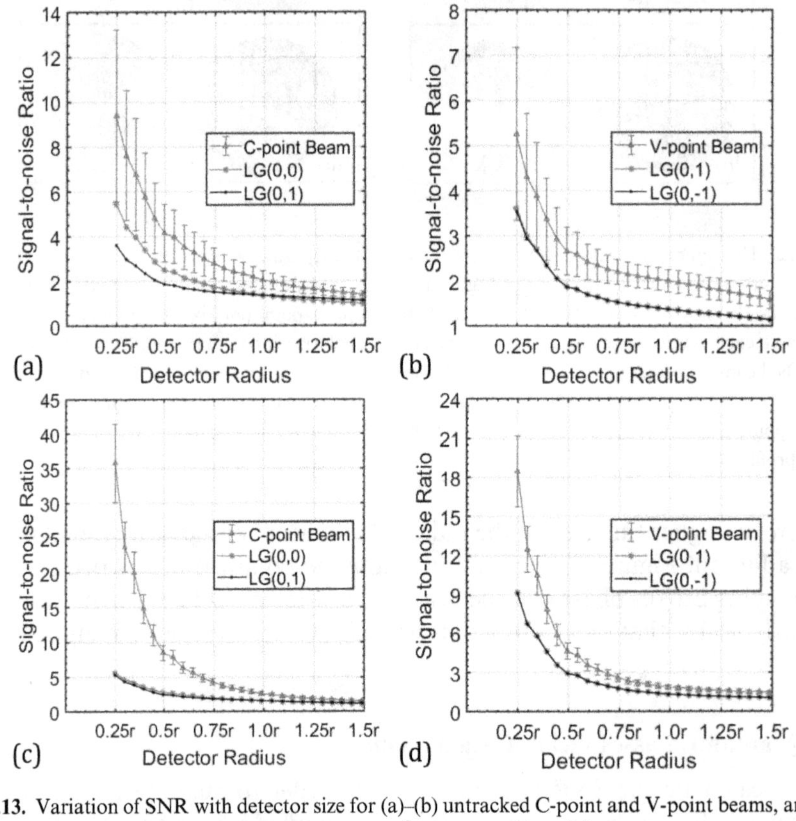

Figure 9.13. Variation of SNR with detector size for (a)–(b) untracked C-point and V-point beams, and (c)–(d) tracked C-point and V-point beams respectively for $C_n^2 = 1 \times 10^{-14}$ m$^{-2/3}$. Adapted with permission from [20] © 2019 Optical Society of America.

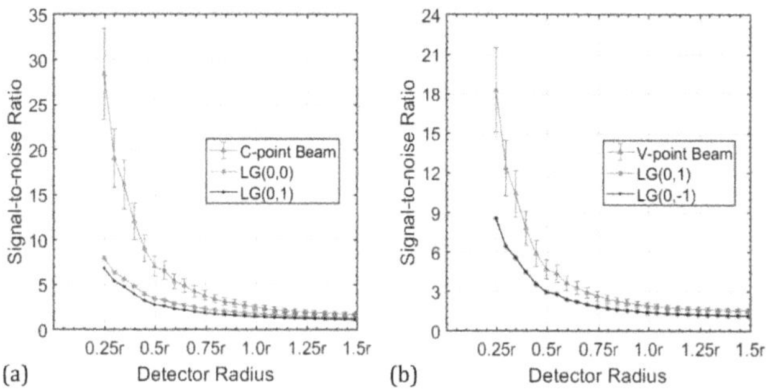

Figure 9.14. Variation of SNR with detector size for tracked (a) C-point and (b) V-point beams for $C_n^2 = 1 \times 10^{-13}$ m$^{-2/3}$. Adapted with permission from [20] © 2019 Optical Society of America.

limited spot size of the Gaussian beam. The SNR values of the two constituting polarizations are also plotted. The error bars in the figures represent the standard deviation of the SNR values calculated over 1000 independent realizations of atmosphere. Each of these 1000 realizations used 20 random phase screens generated using the FFT method with sub-harmonic correction to simulate the effect of turbulence. The error bars are shown on composite vector beam but not on individual polarizations in order to avoid overcrowding of the plot. Figure 9.13 shows that the SNR values for the two polarization singular beams are consistently greater compared to their orthogonal polarization components. At $z = 2$ km, there is not much difference in the SNR values of LG(0,0) and LG(0,1) beams individually, however when added incoherently, a significant improvement in SNR of C-point beam is seen. Similarly, the SNR values of LG(0,1) and LG(0,-1) are almost identical. Adding them incoherently improves the SNR of V-point beam but not as much as in the case of the C-point beam. The SNR values of both C-point and V-point beams show further improvement for the tracked case as compared to the untracked case since now the corresponding beam profile on the detector is better. Figure 9.14 shows the SNR curves for the tracked polarization singular beams (averaged over 1000 realizations) when $C_n^2 = 1 \times 10^{-13}$ m$^{-2/3}$. It can be observed that even under high turbulence, the C-point beam still retains a uniform spot, despite the overall beam experiencing greater spread. The V-point beam does not exhibit this feature under the same realizations of random phase screens. As a result, the C-point beam is able to maintain a higher SNR value compared to the V-point beam or the scalar $l = 0$ and $l = 1$ OAM states.

The correlation coefficient is computed for the tracked polarization singular beams over areas of varying sizes (different radii) at the receiver plane at $z = 2$ km and is presented in table 9.3. The correlation coefficient values for $C_n{}^2 = 1 \times 10^{-13}$ m$^{-2/3}$ follow a similar trend to those for $C_n{}^2 = 1 \times 10^{-14}$ m$^{-2/3}$. However, for the case of high turbulence $C_n{}^2 = 10^{-13}$ m$^{-2/3}$, the beams break up into multiple speckles and experience greater spread, resulting in increased variance in the measured values of ρ_{I_1,I_2}. In general, it is observed that the intensity patterns corresponding to the orthogonal polarization states of the C-point beam have

Table 9.3. Correlation coefficient between the intensity patterns associated with the orthogonal polarization components of C-point and V-point beams at $z = 2$ km for $C_n^2 = 1 \times 10^{-14}\ \mathrm{m}^{-2/3}$ and $C_n^2 = 1 \times 10^{-13}\ \mathrm{m}^{-2/3}$.

Type of Beam	Radius	Value of Correlation Coefficient ρ_{I_1,I_2}	
		$C_n^2 = 1 \times 10^{-14}\ \mathrm{m}^{-2/3}$	$C_n^2 = 1 \times 10^{-13}\ \mathrm{m}^{-2/3}$
C-point	$0.25r$	-0.9514 ± 0.0464	-0.8073 ± 0.2012
	$0.50r$	-0.8767 ± 0.0768	-0.6109 ± 0.2905
	$0.75r$	-0.7305 ± 0.1208	-0.4494 ± 0.3126
	$1.0r$	-0.5371 ± 0.1591	-0.3451 ± 0.3165
V-point	$0.25r$	-0.5050 ± 0.4134	-0.4631 ± 0.4078
	$0.50r$	-0.1448 ± 0.4402	-0.2516 ± 0.4496
	$0.75r$	-0.1341 ± 0.4338	-0.1756 ± 0.4490
	$1.0r$	-0.1951 ± 0.3799	-0.1776 ± 0.3901

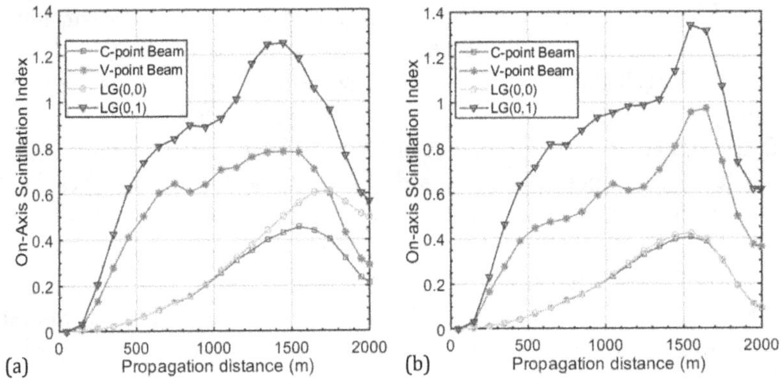

Figure 9.15. Variation of on-axis SI with propagation distance for (a) untracked and (b) tracked beams for $C_n^2 = 1 \times 10^{-14}\ \mathrm{m}^{-2/3}$. The markers represent the location of the random phase screens. Adapted with permission from [20] © 2019 Optical Society of America.

significant negative correlation (correlation coefficient <-0.5) up to a sizable receiver radius for both moderate and high turbulence levels. On the other hand, the orthogonal polarization states of the V-point beam are nearly uncorrelated for the same receiver sizes. The observed results obtained using the multiple phase screen model for turbulence are consistent with those obtained using a single phase screen to model turbulence. The negative intensity correlation in the orthogonal polarization states corresponding to the $l = 0$ and $l = 1$ states is a unique feature of the C-point polarization singularity, which is crucial for robust beam design.

Figure 9.15 shows the variation of the on-axis SI as a function of propagation distance for both untracked or tracked polarization singular beams for $C_n^2 = 1 \times 10^{-14}\ \mathrm{m}^{-2/3}$. The beams have been corrected for beam wander by tracking the centroid of the intensity profile. It is important to note that the tilt caused in the two orthogonal polarization components of the vector beams due to the sub-harmonic terms is of similar nature. Further, it is noticed that the typical beam wander for the simulation parameters was seen to be up to 30% of the free-space diffraction-limited spot size on average.

The traditional measure for assessing beam quality after propagation through turbulence is SI. The on-axis SI is defined as:

$$SI(0, 0, z) = \frac{\langle I(0, 0, z) \rangle}{\sqrt{\langle I(0, 0, z)^2 \rangle - \langle I(0, 0, z) \rangle^2}}. \tag{9.14}$$

Here the ensemble average is taken over a number of realizations of turbulence. In figures 9.15 and 9.16, the on-axis SI values have been calculated over a 3×3 pixel square aperture which is placed on-axis at $(x, y) = (0, 0)$ in the detector plane. The size of this square aperture is below $\sqrt{\lambda L/2\pi}$, which is the required condition for avoiding aperture averaging effects in SI calculation. The plots show the SI values averaged over 1000 turbulence screen realizations. The plots in figure 9.15 show that for both untracked as well as tracked cases, the on-axis SI values of the C-point beam are consistently lower than that for V-point beam. For the tracked case, the on-axis SI for the C-point and LG(0,0) beam appears to be nearly identical over the propagation distances considered here. For a higher atmospheric turbulence case which is represented by $C_n^2 = 1 \times 10^{-13}$ m$^{-2/3}$, the on-axis SI plots for the tracked beams are shown in figure 9.16. It can be seen that the C-point and V-point beams have comparable SI values in the strong turbulence case. The instantaneous SNR curves in figures 9.13 and 9.14, however, show that the SNR at $z = 2$ km for the C-point beam is significantly higher compared to the LG(0,0) beam. In on-axis SI calculations, one looks at the on-axis or central intensity of the beam, therefore, for beams which naturally contain a null intensity at their core, like vortex beams,

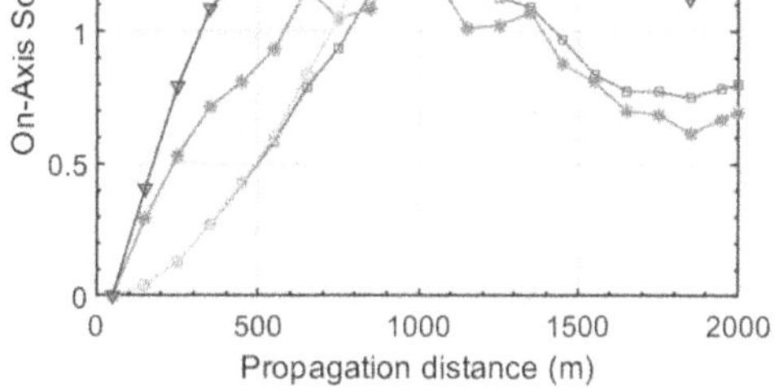

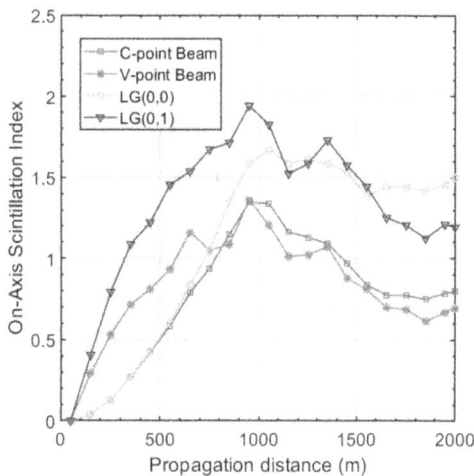

Figure 9.16. Variation of on-axis SI with propagation distance for tracked beams for $C_n{}^2 = 1 \times 10^{-13}$ m$^{-2/3}$. The markers represent the location of the random phase screens

on-axis SI values are very high at low z values. These beams on further propagation through turbulence develop a speckled appearance and experience larger wandering and beam spreading. As the long-averaged beam spot is obtained by summing over all the individual realizations, this process fills up the null at the center for untracked beams resulting in a decrease in SI values. Therefore, on-axis SI might be a biased approach for characterizing fluctuations in vortex beams and on-axis SI alone might not give correct insight into the beam behavior in turbulence. In this respect, measuring instantaneous SNR can be a valuable additional parameter to assess quality of the spatial beam profile.

We have presented a robust beam engineering principle based on the complementary diffraction effect between diffraction patterns associated with $l = 0, 1$ OAM states. The robust beam design idea as presented here may have important implications for adaptive optics as well as FSO communication systems. These topics need to be studied in detail in future.

9.6 Irradiance probability distribution due to engineered beams

In this section, we briefly discuss the effect of beam engineering on the probability distribution function $p(I)$ associated with the on-axis beam irradiance after the beam has propagated through turbulence. The form of $p(I)$ is important for applications in FSO communication. Based on discussion in the prior sections, the engineered beam with a C-point polarization singularity structure is most interesting to us. Other engineered beam structures may be studied in a similar manner. In particular, for the present illustration, we set up the multi-phase screen beam propagation simulation for atmospheric conditions given in table 9.2. We plot histograms of the normalized on-axis irradiance obtained over 2500 atmospheric realizations. The corresponding histograms for scalar Gaussian beam and C-point beam are shown in figure 9.17.

The histograms show that after long-range propagation, for a scalar focused Gaussian beam, the on-axis intensity distribution $p(I)$ is exponential in nature, as

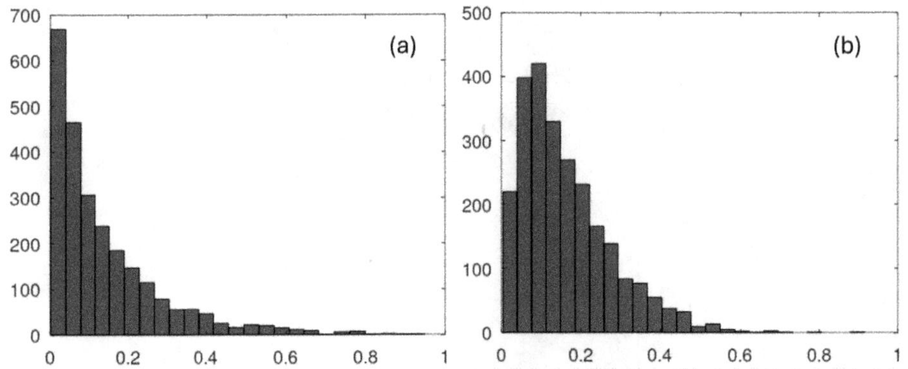

Figure 9.17. Variation of on-axis irradiance distribution for focused (a) scalar Gaussian and (b) engineered (C-point) beam. The plot represents a histogram using 2500 atmospheric realizations for normalized irradiance. The scalar Gaussian beam shows nearly exponential decay behavior while for the engineered beam there is a distinct peak shift.

expected for a speckle pattern. On the other hand, for the engineered beam, there is a distinct peak shift in the $p(I)$. This peak shift resulting from the usage of an engineered beam has important system level implications for applications like FSO communication links. One of the major challenges in deploying FSO systems is the impact of atmospheric turbulence. The exponential behavior of $p(I)$ for a scalar beam as in figure 9.17(a) implies that the low irradiance values have the highest probability. If a detector receives such a null (or smaller than the threshold) irradiance with high probability, there is a significant risk of link failure. The peak shift in $p(I)$ due to an engineered beam suggests that the link failure rate may be mitigated if an engineered beam, rather than a scalar beam, is used in an FSO system. From the standpoint of an FSO system, the hardware changes required are minimal; the primary adjustment needed is the spatial engineering of the transmitted beam. The mechanisms for frequency modulation and demodulation for information transmission may be similar to what is used in the conventional systems. A detailed study of engineered beam-based FSO systems may be warranted in the future to understand the quantitative performance advantages in the context of FSO systems.

9.7 Higher order engineered beams

The analysis of engineered beams presented in this chapter so far has mainly concentrated on use of the two polarization degrees of freedom. From the discussion in section 3.5, we note the description of partially spatially coherent light in terms of the coherent mode representation. The partially coherent scalar light beam was described in the form:

$$u(\mathbf{r}, \nu) = \sum_n a_n \, \phi_n(\mathbf{r}, \nu), \tag{9.15}$$

where $\phi_n(\mathbf{r}, \nu)$ are the eigenfunctions associated with its cross-spectral density $W(\mathbf{r}_1, \mathbf{r}_2, \nu)$, and the coefficients a_n are uncorrelated. We recall that modeling the propagation of partially coherent light through turbulence involves propagating individual coherent modes through the same turbulence conditions using the split-step approach described in chapter 8. The total irradiance may be obtained by adding the contributions from individual coherent modes incoherently with weights given by $|a_n|^2$. The vector beams discussed in this chapter can thus essentially be thought to behave like two-mode partially coherent light. The OAM modes used in the engineered beam design seem to offer diversity in their irradiance after turbulence propagation, as evidenced by negative intensity correlations, at least up to moderate levels of turbulence. The speckle diversity then leads to smoother beam profiles that are a function of the specific beam design used. The results discussed so far in this chapter, suggest that one may use an additional set of coherent modes to get further speckle diversity and thus smooth the overall beam profile.

We emphasize that, in general, the irradiance distribution is much easier to detect compared to the correlation function $W(\mathbf{r}_1, \mathbf{r}_2, \nu)$. In beam engineering applications, defence systems or FSO communication links, one is also interested in maintaining beam directivity in addition to speckle or scintillation reduction. Existing system hardware in these applications can benefit greatly if a smooth instantaneous beam profile can be achieved even after propagation through turbulence. It is well-known in the literature that a fully coherent light (single coherent mode) may show high directivity but high scintillation. Fully spatially incoherent light will have a large number of coherent modes and as a result, the beam irradiance after turbulence may have no speckle but the directivity will be completely lost [49–51]. One may achieve a good trade-off in beam smoothness/quality and beam directivity if a few-mode partially coherent light is used. The number of uncorrelated coherent modes to be used will eventually depend on the requirements of the application at hand and allowable system complexity.

Since polarization can provide only two degrees of freedom as seen with the discussion in the context of vector beams, designs with more than two coherent modes may involve collinear combination of multiple independent laser sources that have the same nominal wavelength but engineered to have distinct spatial mode profiles. Laser beam combination through multiple apertures or use of spectral beam combination by use of diffraction grating are already active areas of research, particularly for high-power laser applications. In figure 9.18, we illustrate typical beam profiles observed

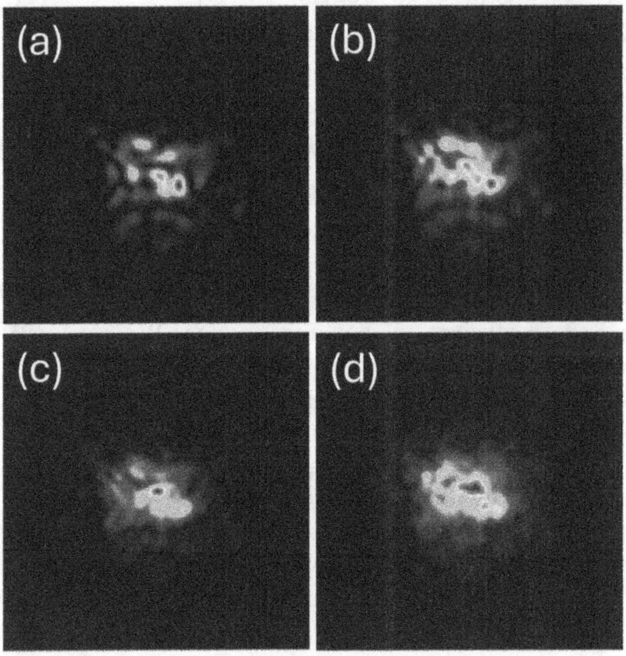

Figure 9.18. Illustration of focused beam irradiance at 2 km for (a) scalar Gaussian beam ($l = 0$) and engineered beams consisting of incoherent combination of (b) $l = 0, 1$, (c) $l = -1, 0, 1$, and (d) $l = -2, -1, 0, 1, 2$ OAM modes. The initial beam waist used is 10 cm for all the beams and $C_n^2 = 5 \times 10^{-14}$ m$^{-2/3}$. The simulation has been performed with the split-step phase screen method, as in section 9.5. Individual OAM modes have equal energy. The images in (a–d) are zoomed-in versions of spot patterns, each covering a dimension of 20 mm × 20 mm.

when multiple collinear OAM modes are combined incoherently, for example, using multiple independent lasers and suitable combiner optics. The simulation clearly illustrates the speckle reduction on use of multiple collinear incoherent modes for beam engineering. These illustrations represent just one realization of the atmosphere. As the atmosphere fluctuates, the scalar beam shown in figure 9.18(a) will show statistically similar irradiance pattern, consisting of multiple separated spots. On the other hand, the engineered beams depicted in figures 9.18(b)–(d) will progressively show a more continuous beam without significant nulls. In this case, if the resultant beam is incident on a detector or target at the receiving end, the energy delivered across the beam spot remains more uniform, even if the atmosphere is fluctuating over time. Such higher-order engineered beams require further detailed study, as they may provide significant practical advantages.

References

[1] Baykal Y, Eyyuboglu H T and Cai Y 2009 Effect of beam types on the scintillations: a review *Atmospheric Propagation of Electromagnetic Waves* **vol 7200** ed O Korotkova (Bellingham, WA: International Society for Optics and Photonics, SPIE) pp 7–21

[2] Eyyuboglu H T 2011 Annular, cosh and cos Gaussian beams in strong turbulence *Appl. Phys.* B **103** 763–9 6

[3] Cai Y 2006 Propagation of various flat-topped beams in a turbulent atmosphere *J. Opt. A: Pure Appl. Opt.* **8** 537–45 4

[4] Cai Y and He S 2006 Propagation of various dark hollow beams in a turbulent atmosphere *Opt. Express* **14** 1353–67 2

[5] Korotkova O 2008 Scintillation index of a stochastic electromagnetic beam propagating in random media *Opt. Commun.* **281** 2342–8

[6] Gbur G and Wolf E 2002 Spreading of partially coherent beams in random media *J. Opt. Soc. Am.* A **19** 1592–8 8

[7] Borah D K and Voelz D G 2010 Spatially partially coherent beam parameter optimization for free space optical communications *Opt. Express* **18** 20746–58 9

[8] Ricklin J C and Davidson F M 2002 Atmospheric turbulence effects on a partially coherent Gaussian beam: implications for free-space laser communication *J. Opt. Soc. Am.* A **19** 1794–802 9

[9] Lochab P, Senthilkumaran P and Khare K 2017 Robust laser beam engineering using polarization and angular momentum diversity *Opt. Express* **25** 17524–9 9

[10] Wheelon A D 2001 *Electromagnetic Scintillation* **vol 2** (Cambridge: Cambridge University Press)

[11] Wei C, Wu D, Liang C, Wang F and Cai Y 2015 Experimental verification of significant reduction of turbulence-induced scintillation in a full Poincaré beam *Opt. Express* **23** 24331–41 9

[12] Lochab P, Senthilkumaran P and Khare K 2018 Designer vector beams maintaining a robust intensity profile on propagation through turbulence *Phys. Rev.* A **98** 023831 8

[13] Cheng W, Haus J W and Zhan Q 2009 Propagation of vector vortex beams through a turbulent atmosphere *Opt. Express* **17** 17829–36 9

[14] Chen Z, Cui S, Zhang L, Sun C, Xiong M and Pu J 2014 Measuring the intensity fluctuation of partially coherent radially polarized beams in atmospheric turbulence *Opt. Express* **22** 18278–83

[15] Cox M A, Rosales-Guzmán C, Lavery M J, Versfeld D J and Forbes A 2016 On the resilience of scalar and vector vortex modes in turbulence *Opt. Express* **24** 18105–13 8

[16] Gu Y, Korotkova O and Gbur G 2009 Scintillation of nonuniformly polarized beams in atmospheric turbulence *Opt. Lett.* **34** 2261–3

[17] Gu Y and Gbur G 2012 Reduction of turbulence-induced scintillation by nonuniformly polarized beam arrays *Opt. Lett.* **37** 1553–5

[18] Goodman J W 1975 *Statistical Properties of Laser Speckle Patterns* ed J C Dainty (Berlin: Springer)

[19] Song Y, Milam D and Hill W T 1999 Long, narrow all-light atom guide *Opt. Lett.* **24** 1805–7 12

[20] Lochab P P, Senthilkumaran P and Khare K 2019 Propapation of converging polarization singular beams through atmospheric turbulence *Appl. Opt.* **58** 6335–45

[21] Dickson L D 1970 Characteristics of a propagating Gaussian beam *Appl. Opt.* **9** 1854–61 8

[22] Andrews L C, Phillips R L and Weeks A R 1997 Propagation of a Gaussian-beam wave through a random phase screen *Waves Random Media* **7** 229–44

[23] Ricklin J C, Miller W B and Andrews L C 1995 Effective beam parameters and the turbulent beam waist for convergent Gaussian beams *Appl. Opt.* **34** 7059–65 10

[24] Dios F, Rubio J A, Rodriguez A and Comeron A 2004 Scintillation and beam-wander analysis in an optical ground station-satellite uplink *Appl. Opt.* **43** 3866–73 7

[25] Recolons J, Andrews L C and Phillips R L 2007 Analysis of beam wander effects for a horizontal-path propagating Gaussian-beam wave: focused beam case *Opt. Eng.* **46** 1–11

[26] Dowling J A and Livingston P M 1973 Behavior of focused beams in atmospheric turbulence: measurements and comments on the theory *J. Opt. Soc. Am.* **63** 846–58 7

[27] Birch P, Ituen I, Young R and Chatwin C 2015 Long-distance bessel beam propagation through Kolmogorov turbulence *J. Opt. Soc. Am.* A **32** 2066–73 11

[28] Bouchal Z 2002 Resistance of nondiffracting vortex beam against amplitude and phase perturbations *Opt. Commun.* **210** 155–64

[29] Gbur G and Tyson R K 2008 Vortex beam propagation through atmospheric turbulence and topological charge conservation *J. Opt. Soc. Am.* A **25** 225–30 1

[30] Aksenov V P and Kolosov V V 2015 Scintillations of optical vortex in randomly inhomogeneous medium *Photon. Res.* **3** 44–7 4

[31] Eyyuboğlu H T 2016 Scintillation behaviour of vortex beams in strong turbulence region *J. Mod. Opt.* **63** 2374–81

[32] Liu X and Pu J 2011 Investigation on the scintillation reduction of elliptical vortex beams propagating in atmospheric turbulence *Opt. Express* **19** 26444–50 12

[33] Paterson C 2005 Atmospheric turbulence and orbital angular momentum of single photons for optical communication *Phys. Rev. Lett.* **94** 153901 4

[34] Goyal S K, Ibrahim A H, Roux F S, Konrad T and Forbes A 2016 The effect of turbulence on entanglement-based free-space quantum key distribution with photonic orbital angular momentum *J. Opt.* **18** 064002 4

[35] Karimi E, Marrucci L, de Lisio C and Santamato E 2012 Time-division multiplexing of the orbital angular momentum of light *Opt. Lett.* **37** 127–9

[36] Leonhard N, Sorelli G, Shatokhin V N, Reinlein C and Andreas B 2018 Protecting the entanglement of twisted photons by adaptive optics *Phys. Rev.* A **97** 012321 1

[37] Schulz T J 2005 Optimal beams for propagation through random media *Opt. Lett.* **30** 1093–5 5

[38] Voelz D G and Xiao X 2009 Metric for optimizing spatially partially coherent beams for propagation through turbulence *Opt. Eng.* **48** 1–7
[39] Lochab P, Senthilkumaran P and Khare K 2016 Near-core structure of a propagating optical vortex *J. Opt. Soc. Am.* A **33** 2485–90
[40] Tao R, Wang X, Si L, Zhou P and Liu Z 2013 Propagation of focused vector laser beams in turbulent atmosphere *Opt. Laser Technol.* **54** 62–7
[41] Andrews L C and Phillips R L 2005 *Laser Beam Propagation through Random Media* (Bellingham, WA: SPIE Press Book)
[42] Kogelnik H 1965 Imaging of optical modes—resonators with internal lenses *Bell Syst. Tech. J.* **44** 455–94 3
[43] Beckley A M, Brown T G and Alonso M A 2010 Full Poincaré beams *Opt. Express* **18** 10777–85 5
[44] Gibson C J, Bevington P, Oppo G L and Yao A M 2018 Control of polarization rotation in nonlinear propagation of fully structured light *Phys. Rev.* A **97** 033832 3
[45] Churnside J H and Lataitis R J 1990 Wander of an optical beam in the turbulent atmosphere *Appl. Opt.* **29** 926–30 3
[46] Banakh V A, Krekov G M, Mironov V L, Khmelevtsov S S and Tsvik R S 1974 Focused-laser-beam scintillations in the turbulent atmosphere *J. Opt. Soc. Am.* **64** 516–8 4
[47] Titterton P J 1973 Scintillation and transmitter-aperture averaging over vertical paths *J. Opt. Soc. Am.* **63** 439–44 4
[48] Kerr J R and Eiss R 1972 Transmitter-size and focus effects on scintillations *J. Opt. Soc. Am.* **62** 682–4 4
[49] Schulz T J 2005 Optimal beams for propagation through random media *Opt. Lett.* **30** 1093–5
[50] Gbur G 2014 Partially coherent beam propagation in atmospheric turbulence *J. Opt. Soc. Am.* A **31** 2038–45
[51] Nair A, Li Q and Stechmann S N 2023 Scintillation minimization versus intensity maximization in optimal beams *Opt. Lett.* **48** 3865–8

IOP Publishing

Orbital Angular Momentum States of Light (Second Edition)

Propagation through atmospheric turbulence

Kedar Khare, Priyanka Lochab and Paramasivam Senthilkumaran

Chapter 10

Speckle in structured light with applications

This chapter opens a new line of investigation into the age-old speckle phenomena using structured light beams, which we introduced in the previous chapter. While speckle in scalar (or homogeneously polarized) illumination has been well studied, the use of structured light throws up some surprises. In particular, the speckle intensity patterns are seen to have unique textures when structured beams with different designs pass through an identical random medium. Some practical potential applications of the speckle texture analysis are presented as well.

10.1 Speckle phenomena in optics

Speckle refers to a complex interference phenomenon associated with coherent light illumination and has been studied extensively since the invention of laser [1, 2]. When a laser beam reflects off a rough surface like a wall, the resultant intensity pattern on propagation exhibits fine-scale granular features that are referred to as speckle. The majority of natural or artificial materials encountered in the real world exhibit roughness at scales comparable to optical wavelengths, thus randomly perturbing the spatial phase profile of the reflected light beam. On an observation screen (see figure 10.1), these random phase perturbations lead to a complex interference with granular intensity pattern. Speckle is also observed for the same reason when a coherent light beam passes through a ground glass diffuser or other random media such as Earth's atmosphere or the vitreous humor in the human eye. While laser sources offer efficient directed energy beams, the appearance of speckle can limit their direct usage in applications like microscopy, large-scale displays, etc. Considerable efforts are therefore required to mitigate the speckle [3]. In this sense, speckle has traditionally been considered as an undesirable effect. On the other hand, speckle has also found multiple applications in computational sensing and imaging systems, thus enabling novel 3D imaging concepts, compact spectrometers, and ability to image through scattering media by means of correlation imaging [4].

doi:10.1088/978-0-7503-5959-7ch10

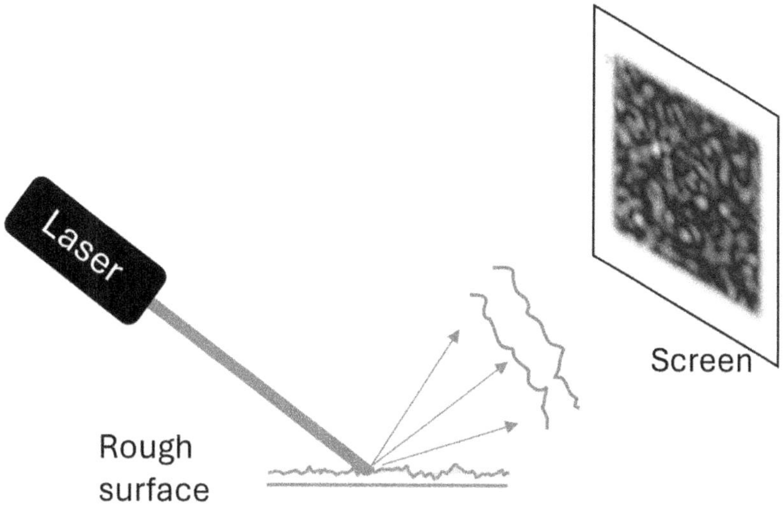

Figure 10.1. Generation of speckles from reflection of spatially coherent light beam from a rough surface.

Our aim in this chapter is to briefly review the statistics of speckle in scalar as well as vector beams. While speckle has been studied for several decades, most of the speckle statistics studies are focused on point statistics with scalar (or homogeneously polarized) illumination. The space and wavelength dependence of speckle statistics has also been studied in detail [5]. Our goal is to highlight some interesting spatial textural properties that are observed in speckle intensity patterns with vector beam illumination. We describe results of some recent investigations that open up several exciting possibilities for new investigations in the future. The individual bright regions in a speckle are known to have a transverse dimension

$$d_{\text{transverse}} \approx \frac{\lambda z}{D}, \tag{10.1}$$

and a longitudinal dimension

$$d_{\text{longitudinal}} = \frac{\lambda z^2}{D^2}. \tag{10.2}$$

Here λ is the illuminating wavelength, z is the distance between the rough surface and the observation screen and D is the dimension of the illuminating laser spot. The expressions for $d_{\text{transverse}}$ and $d_{\text{longitudinal}}$ above suggest that the speckles have a feature size dependent on the diffraction-limit of the experimental configuration. These expressions can be obtained by evaluating the correlation $\langle u^*(\mathbf{r}_1)\, u(\mathbf{r}_2)\rangle$ of the field $u(\mathbf{r})$ at two spatially separated points $\mathbf{r}_1$ and $\mathbf{r}_2$.

10.2 Intensity and phase statistics of scalar speckle

Referring to figure 10.1, we note that the scalar field at any point in the observation screen has contributions from a large number of scattering centers on the rough reflecting surface. Further, their contributions are expected to have a randomly varying phase which is distributed uniformly over $[0, 2\pi]$, if the reflecting surface has fine scale variations on the wavelength scale. The total scalar field at a given observation point can therefore be expressed as

$$u(\mathbf{r}) = \sum_n a_n \exp(i\theta_n). \tag{10.3}$$

Here a_n and θ_n are amplitude and phase of the contribution due to individual scattering centers at the observation point $\vec{r}$. We assume monochromatic illumination of wavelength λ and an $\exp(-i\omega t)$ time dependence, which has been omitted throughout for concise notation. The summation in equation (10.3) is over a very large number of scatterers on the rough surface. As per the central limit theorem [6], we therefore expect the scalar field $u(\mathbf{r})$ to be described by a complex circular Gaussian random process. We will denote the real and imaginary parts of $u(\mathbf{r})$ as $u_r(\mathbf{r})$ and $u_i(\mathbf{r})$, respectively. The variables $u_r(\mathbf{r})$ and $u_i(\mathbf{r})$ are expected to be statistically independent and therefore their joint probability density function is expected to have the form

$$p(u_r, u_i) = \frac{1}{2\pi\sigma^2} \exp\left[-\frac{(u_r^2 + u_i^2)}{2\sigma^2}\right]. \tag{10.4}$$

Using the polar coordinate transformation $A = \sqrt{u_r^2 + u_i^2}$ and $\theta = \arctan(u_i/u_r)$, we have

$$p(u_r, u_i)\, du_r\, du_i = p(A, \theta) dA d\theta. \tag{10.5}$$

Further, since $du_r du_i = A dA d\theta$, we have

$$p(A, \theta) = \frac{A}{2\pi\sigma^2} \exp\left(-\frac{A^2}{2\sigma^2}\right). \tag{10.6}$$

Since u_r and u_i are statistically independent, the phase $\theta(\mathbf{r})$ of $u(\mathbf{r})$ is uniformly distributed in $[0, 2\pi]$. Therefore, integrating over 2π, we get the probability distribution for the scalar field amplitude,

$$p(A) = \frac{A}{\sigma^2} \exp\left(-\frac{A^2}{2\sigma^2}\right), \tag{10.7}$$

which is known as the Rayleigh distribution. By making further transformation $I = A^2$ and applying the probability density transformation law, we have

$$p(I) = \frac{1}{I_0} \exp\left(-\frac{I}{I_0}\right). \tag{10.8}$$

Here $I_0 = 2\sigma^2$ and $p(I)$ is the well known exponential probability density function associated with speckle intensity statistics [1, 2]. Experimentally, such a distribution is observed readily by performing ensemble average using a detector whose area of integration is much smaller than typical speckle size. In order to perform ensemble average, different parts of the rough surface need to brought into the illuminating beam while observing the intensity at a fixed location **r**.

When two independent speckle patterns are added incoherently as in case of polarization speckle [7], the resultant intensity statistics may be obtained by convolution of $p(I)$ in equation (10.8) with itself. The polarization speckle may for example be generated by separating a 45-degree polarized laser into x and y polarization components, passing these individual polarization components through two diffusers and then combining the two scattered beams back incoherently. The point statistics of the sum of two independent speckles with the same intensity statistics is given by

$$p(I = I_1 + I_2) = \frac{I}{I_0}\exp\left(-\frac{I}{I_0}\right). \tag{10.9}$$

It is important to note that the exponential intensity distribution in equation (10.8) peaks at zero intensity, whereas the peak of the intensity distribution in equation (10.9) is shifted to a non-zero intensity value. The reason for this shift is that two independent speckles do not likely have isolated intensity nulls at identical spatial location. While sums of independent speckle patterns and their statistics has been studied in detail in prior literature, our focus here is on correlated speckles produced by two orbital angular momentum (OAM) states, as we will discuss in the next section. In the previous chapter 9, we have already noted the negative intensity correlation between diffraction patterns produced by the $l = 0$ and $l = 1$ OAM states when they see the same random atmosphere. Here we extend that observation to a more general case of diffraction of polarization structured beams by random diffuser screens.

10.3 Speckle in polarization structured light

In this section, we discuss some recent observations regarding speckle in the polarization structured light beams having the V-point and C-point polarization singularities [8]. The schematic setup for observing speckle in polarization structured light is depicted in figure 10.2(a). A polarization singular beam generator is essentially one of the arrangements discussed in chapter 6. The illumination polarization states to be studied include homogeneous linear polarization, and the V-point and C-point polarization singularities, as illustrated in figures 10.2(b)–(d), respectively. The homogeneous linear polarization case is the same as has been traditionally studied in speckle literature. The input beam is passed through a convex lens, a diffuser glass, and the speckle intensity is recorded in the back focal plane of the lens on an array detector. The aperture in the beam path near the lens is used to control the speckle size on the camera sensor. As described in detail in chapter 6, the transverse E-field containing polarization singularities (figures 10.2(c) and (d)) that is used for illumination may be described as

$$\begin{aligned}\mathbf{E}(r, \theta) &= [\hat{e}_1 E_1(r, \theta) + \hat{e}_2 E_2(r, \theta)] \\ &= [\hat{e}_1 r^{|l_1|}\exp(il_1\theta) + \hat{e}_2 r^{|l_2|}\exp(il_2\theta)]\psi(r, \theta).\end{aligned} \tag{10.10}$$

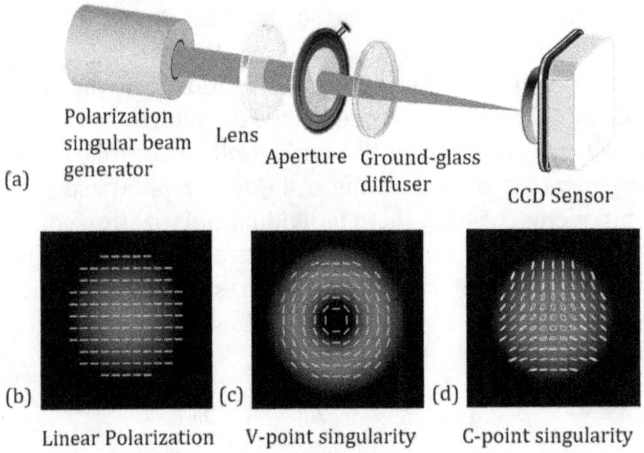

Figure 10.2. Schematic setup for observing speckle in polarization structured light. Adapted with permission from [8] ©Taylor and Francis.

Here $E_1(r, \theta)$ and $E_2(r, \theta)$ represent two orthogonally polarized components associated with unit vectors $\hat{e}_1$ and $\hat{e}_2$, denoting circular basis right circular polarization, left circular polarization, respectively, and $\psi(r, \theta)$ is the amplitude profile of the native Gaussian beam. The two polarization components $E_1(r, \theta)$ and $E_2(r, \theta)$ can be normalized to have the same energy. We have not shown the normalization factor in equation (10.10). The homogeneous polarization state in figure 10.2(b) corresponds to $l_1 = l_2 = 0$, while the inhomogeneous polarization cases shown in figure 10.2(c) and (d) correspond to the lowest order polarization singularities. In particular, the V-point singularity corresponds to the case $l_1 = -l_2 = 1$ and the C-point singularity corresponds to the case $l_1 = 0$ and $l_2 = 1$. The important point to note here is that, unlike the case of polarization speckle [1], the two orthogonal polarization components $E_1(r, \theta)$ and $E_2(r, \theta)$ of the incident field now see the same portion of the diffuser glass. Assuming negligible polarization cross-talk, the experimental arrangement essentially records two orthogonally polarized Fraunhofer diffraction patterns that add incoherently in intensity. Denoting the aperture function by $A_0(r, \theta)$ and the transmission of ground glass diffuser by $t(r, \theta)$, the irradiance recorded on camera may be represented as:

$$I_{\text{camera}}(\rho, \phi) = |\mathcal{F}\{t(r, \theta)A_0(r, \theta)E_1(r, \theta)\}|^2 + |\mathcal{F}\{t(r, \theta)A_0(r, \theta)E_2(r, \theta)\}|^2. \quad (10.11)$$

This seemingly benign expression has some surprises in store when examined pictorially. In figure 10.3(a)–(c) we observe speckle patterns due to $l = 0, 1, -1$ OAM state illuminations, respectively. At first glance, these images do not have much interesting visual information. However, these speckle patterns have been generated by the identical scattering centers by different illuminating beams. Therefore, they are expected to have some implicit correlation, which is evident when the speckle patterns due to $l = 0, 1$ (C-point illumination) and $l = 1, -1$ (V-point illumination) OAM states are added incoherently, as in figures 10.3(d) and (e), respectively. We observe that the speckle pattern in figure 10.3(d) has a sponge-like

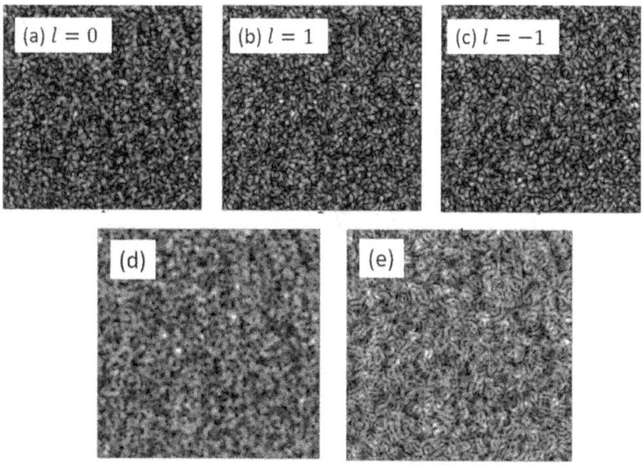

Figure 10.3. Speckle in polarization structured illumination. Scalar speckle patterns with (a) $l = 0$, (b) $l = 1$ and $l = -1$ OAM states illumination. The speckle pattern (d) is obtained by addition of (a), (b) and shows 'sponge'-like speckle texture that we refer to as sponge-speckle or Ceckle. The speckle pattern (e) is obtained by addition of (b), (c) and shows 'noodle'-like speckle texture that we refer to as noodle-speckle or Veckle. Adapted with permission from [8] ©Taylor and Francis.

texture while the speckle pattern in figure 10.3(e) has a noodle-like texture [8]. The distinct textures arise out of different incoherent combinations of speckles from various OAM state illuminations. The texture properties of speckle in polarization structured illumination are an interesting phenomenon suggesting that there is more to speckle statistics beyond the usual point intensity and phase distribution. We coin new names 'Ceckle' and 'Veckle' to refer to speckle patterns resulting out of illuminations containing C-point and V-point polarization singularities. The first letters C and V indicate the nature of illumination. Coincidentally, the 'S' in the word speckle may be thought to be associated with the scalar illumination used.

It may be emphasized that when the diffuser screen is moved, thus bringing new scatterers into the illumination, the speckle patterns observed do change in detail as expected. Their sponge or noodle texture character, however, remains conserved suggesting that the texture properties are a characteristic of the nature of polarization structured illumination. For comparison, we evaluated the ensemble-averaged point intensity statistics $p(I)$ with 1000 realizations of random diffuser screen. The corresponding histogram of observed intensities at a fixed (on-axis) point are shown as histograms in figure 10.4. The histogram for scalar illumination in figure 10.4(a) shows the expected exponential behavior. Interestingly, the point intensity histograms for the Ceckle and Veckle cases as in figures 10.4(b) and (c) are quite similar to each other despite the fact that their texture is clearly distinct. This observation suggests that the point intensity statistics which has been used traditionally is not a good measure to capture the texture properties of the speckle patterns in polarization structured illumination. We therefore need new methodologies to represent the textural properties. Luckily, the image processing literature

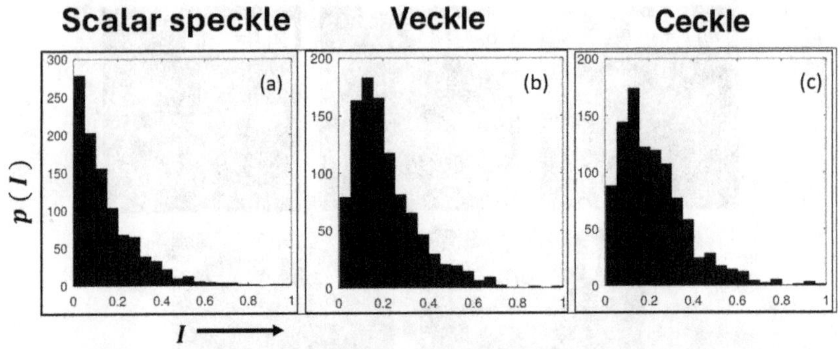

Figure 10.4. Point intensity histograms in speckle with (a) scalar ($l = 0$) illumination, (b) V-point illumination and (c) C-point illumination. Adapted with permission from [8] © Taylor and Francis.

Figure 10.5. Illustration of various image textures.

already has such a measure—the Gray-Level Co-occurrence Matrix (GLCM)—which has been known for several decades. In the next section we describe the GLCM approach to distinguishing texture in images.

10.4 Texture classification: gray-level co-occurence matrix

Textural features are fundamental to image classification and segmentation in computer vision, pattern recognition, object detection, facial recognition, surface defect detection, satellite imagery analysis, and medical imaging for identifying anomalies like tumors and lesions [9–17]. In this section, we will discuss the basics of texture-based classification using GLCMs.

An image encompasses both tonal and textural elements. Figure 10.5 gives an illustration of textural differences in common day-to-day images, where regions of different colors represent different intensity values. These various hues represent the tonal variations in the image, while the spatial or statistical distribution of these various shades represents the texture. As seen in figure 10.5, the human eye can easily detect textural patterns, as coarse, fine, bumpy or grainy. However, mathematically quantifying the texture precisely is very complex.

Haralick *et al* [10, 11] have identified several key textural features that can be extracted from an image, which are highly effective for distinguishing between various types of image data. They have introduced a set of gray-tone spatial-dependence probability-distribution matrices, known as GLCMs, which can be used to derive significant statistical textural characteristics from images. These matrices are based on the concept that texture information in an image is encoded in the spatial relationships among the gray levels present. Essentially, a GLCM is a matrix that captures the frequency or probability (P_{ij}) of different gray-level combinations occurring together within the image. Specifically, it represents the frequency of neighboring pixels, separated by a distance d, where one pixel has a gray level of i and the adjacent pixel has a gray level of j. In short, a GLCM is a matrix that records how often different combinations of pixel brightness values (gray levels) occur together in an image.

Let us understand the calculation of GLCM using a simple example, as shown in figure 10.6. Here, the raw image with the corresponding pixel gray levels in shown in

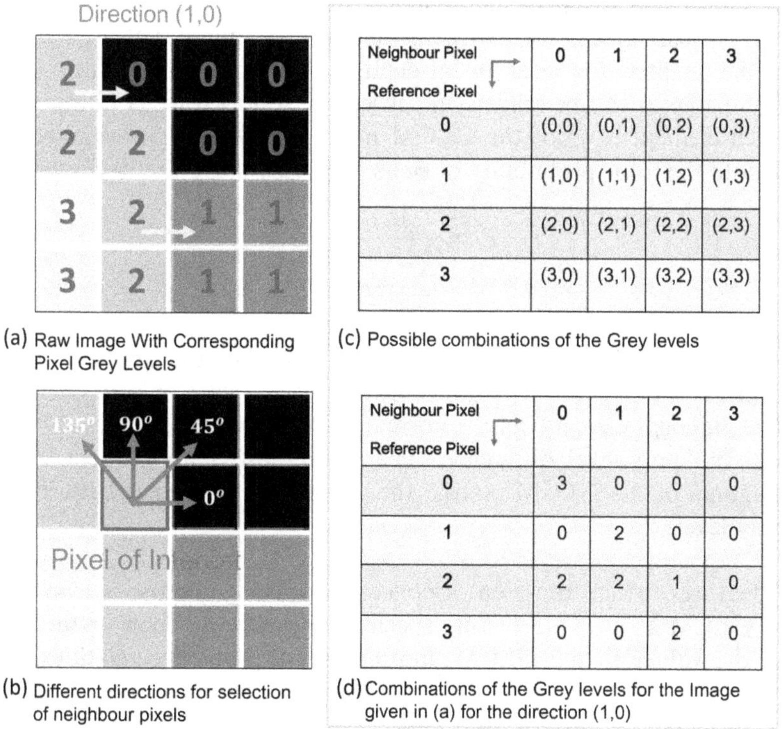

Neighbour Pixel / Reference Pixel	0	1	2	3
0	(0,0)	(0,1)	(0,2)	(0,3)
1	(1,0)	(1,1)	(1,2)	(1,3)
2	(2,0)	(2,1)	(2,2)	(2,3)
3	(3,0)	(3,1)	(3,2)	(3,3)

Neighbour Pixel / Reference Pixel	0	1	2	3
0	3	0	0	0
1	0	2	0	0
2	2	2	1	0
3	0	0	2	0

Figure 10.6. (a) A simple 4×4 image with 4 gray levels is shown. (b) Different directions for selection of neighbor pixels. (c) Possible nearest neighbor combinations of gray levels. (d) Evaluation of the combinations of gray levels for the image shown in (a), along the (1, 0) direction.

(a). For determination of the GLCM matrix of this image, we need to examine the relationship between pairs of pixels, known as reference and neighbor pixels. The neighboring pixels for GLCM calculation can be chosen along any angle (θ) in the image, where horizontally aligned pixels are said to be at 0°, while vertically aligned pixels are at 90°. Figure 10.6(b) shows some other angular nearest neighbors that can also be chosen like 45° or 135°. In the current example, let us choose the neighbor pixel to be the one directly to the right of each reference pixel. This can be represented as the direction (1, 0), as shown by yellow arrows in figure 10.6(a). This means that for calculating different possible gray-level combinations in this direction, we will consider one pixel over in the x-direction and none in the y-direction. For example, a reference pixel of 2 and its neighbor of 1 would contribute one count to the matrix element (2, 1) and one count to the matrix element (1, 2). Then the resulting possible gray-level combinations are shown in figure 10.6(c). Each pixel of the image is considered as the reference pixel sequentially, starting from the upper left corner and moving towards the lower right. Pixels on the right edge lack a right-hand neighbor and will therefore be excluded from this count. In general, the separation distance d between pixel of interest and its neighbor can be chosen as per need for different types of datasets. The separation distance d should best represent the scale of correlation present in the dataset. The combinations of gray levels for the given image along the (1, 0) direction are tabulated in figure 10.6(d). However, the GLCM matrix needs to satisfy the following properties

1. It is a square symmetric matrix having the same number of rows and columns as the quantization level (or bit depth) of the image. The matrix obtained in figure 10.6(d) can be made symmetrical by adding it to its transpose.
2. Each element (i, j) of the GLCM matrix needs to be normalized so as to denote the joint probability of pairs of pixels, using the below equation

$$P_{ij} = \frac{M_{i,j}}{\sum\limits_{i,j=0}^{N-1} M_{i,j}} \tag{10.12}$$

   where $M_{i,j}$ denotes the value in the (i, j) cell of the above-obtained symmetrical GLCM and N denotes the number of rows or columns.
3. The diagonal elements of the normalized GLCM, thus represents pixel pairs with no gray-level difference. The further away an element is from the diagonal of the GLCM matrix, the greater the difference between the pixel gray levels.

Textural features, which represent a collection of image statistics, can be derived from the GLCM to provide texture-specific information. Each texture feature is based on the values in the GLCM and assesses a particular relationship among neighboring pixels in the image. Textural features are sensitive to specific texture elements such as fine, coarse, smooth, rippled, mottled, irregular, or lineated patterns, offering unique insights into the image texture's characteristics. Some key second-order statistical features, or classifiers, that can be derived from the

Autocorrelation	Homogeneity	Contrast	Dissimilarity
Cluster Shade	Maximum Probability	Entropy	Inverse Difference
Sum Variance	Correlation	Energy	Variance
Sum Average	Difference Variance	Difference Entropy	Cluster Prominence
Information Measure of Correlation	Sum Entropy	Inverse Difference Normalized	Inverse Moment Normalized

Figure 10.7. Numerical parameters derived from GLCM analysis.

GLCM include (a) contrast, (b) correlation, (c) homogeneity, and (d) energy [18]. The mathematical definitions of these features are provided in [11]. Contrast measures the regional fluctuations within the GLCM, while correlation quantifies the relationship between a pixel and its neighboring pixel over the entire image. Homogeneity assesses how close the distribution of GLCM elements is to the diagonal, and energy, also known as angular second moment, calculates the sum of squared elements in the GLCM, reflecting the uniformity of the image. A total of 20 such features, as listed in figure 10.7, can be extracted from the GLCM analysis. The complete mathematical definitions of these parameters are available in the standard literature of texture analysis. These texture parameters can serve as input parameters or classifiers to various machine learning methods. An immediate application of the GLCM methodology involves its use in classification of speckle texture. In the next section, we will explore the valuable insights that principle component analysis (PCA) based texture analysis of speckle patterns can offer, including the identification of polarization singular beams and the characterization of atmospheric turbulence.

10.5 Novel applications of speckle texture

With the understanding of GLCM-based texture analysis, we are now ready to revisit the challenge of identifying polarization structured illumination by examining the textural properties of speckle patterns. The computation of the GLCM and the extraction of its parameters are typically carried out using fast computational algorithms available in standard programming tools such as MATLAB [19]. The data comprising experimental and simulation data for C-point and V-point illumination cases is used to obtain the GLCM-based texture classifiers which are later fed as input to a PCA model. Since we saw in figure 10.3 that the speckles do not have any preferred orientation, adding many spatial directions is not necessary for the GLCM formulation, and thus we discuss the results with using only horizontal and 45-degree directions for GLCM calculation. To highlight the significance of speckle texture in both scalar and vector beam illuminations, we also include the speckles obtained using Gaussian beam illumination in the analysis.

Figure 10.8 shows the plot of the first two PCA components obtained using the scalar and vector beam illuminations. The first two PCA components clearly

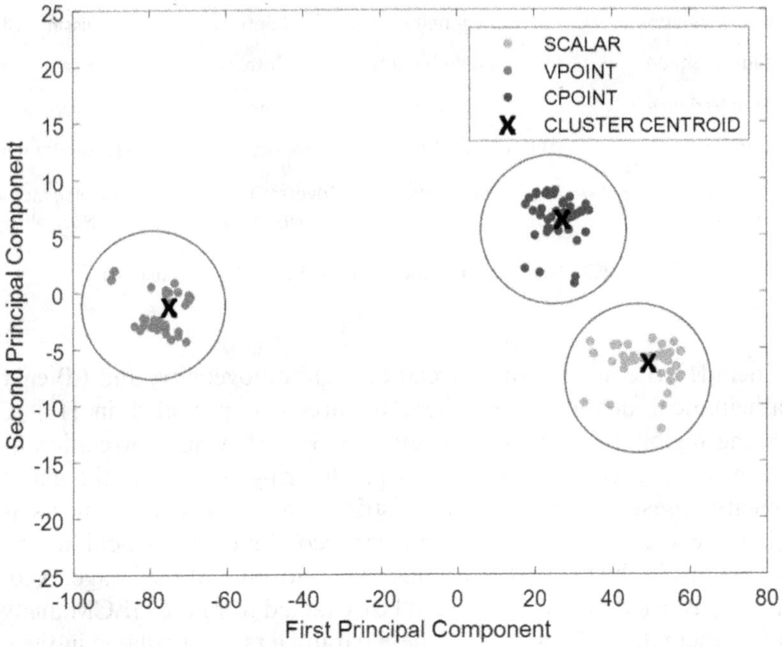

Figure 10.8. PCA plot for GLCM-based texture analysis of speckle obtained with polarization structured light (as illustrated in figure 10.3). Adapted with permission from [8] ©Taylor and Francis.

demonstrate clustering of speckle frames based on whether the illumination is scalar or structured (C-point or V-point), confirming the visually noticeable textural differences between them. The eigenvalues from the PCA analysis indicate that the first two PCA components account for more than 99% of the variance in the data. We can also observe that the speckles from the V-point illumination are quite distinct from those in the Gaussian and C-point beam. The three classes can be further identified using the k-means clustering algorithm, which is readily available in the MATLAB programming environment [20].

The ease with which PCA analysis of textural parameters can differentiate speckles based on their input illumination state, opens up new avenues for employing speckle texture as a communication tool through random media in the future. One can effectively leverage the speckle texture diversity of polarization singular beams as a method for transmitting information through random media such as atmospheric turbulence or biological tissues. For example, alternating between various input polarization states (such as C-point and V-point) can encode binary information, represented by the labels 0 and 1. Even if the transmitted beam results in a fully developed speckle pattern, the texture of this pattern will still be closely linked to the initial polarization state. Consequently, by examining the speckle texture after it has passed through the random medium, one can infer the original polarization state. Additionally, the textural characteristics of a speckle pattern can remain consistent even through a changing random medium such as a moving diffuser or dynamically changing atmospheric turbulence, if the speckle images are captured with a short exposure time. Directly measuring the speckle pattern with an array sensor is also more straightforward than assessing the phase or polarization structure of the speckle field. As a result, communication based on speckle texture could prove to be a novel reliable method in future.

Next we test the speckle texture diversity-based PCA classification for identification of OAM beams after long-range atmospheric propagation. The datasets for focused OAM modes $l = 0$ and $l = 1$ are generated using random phase screen method with modified von Kármán spectrum (as described in chapter 7) with $C_n^2 = 5 \times 10^{-14}\ \mathrm{m}^{-2/3}$ and inner and outer scales, $l_o = 1$ cm and $L_o = 3$ m, respectively. The focused OAM states at $z = 0$ input plane are represented as,

$$LG(0,\ 0) = \sqrt{\frac{2}{\pi W_o^2}} \exp\left(\frac{-r^2}{W_o^2} - \frac{i\pi r^2}{\lambda F_o}\right) \tag{10.13}$$

$$LG(0,\ 1) = \frac{2r}{\sqrt{\pi} W_o^2} \exp(i\theta) \exp\left(\frac{-r^2}{W_o^2} - \frac{i\pi r^2}{\lambda F_o}\right) \tag{10.14}$$

Here $\lambda = 1.55\ \mu$m denotes the wavelength of the light, $W_o = 7$ cm is the beam waist of $l = 0$ state, and $F_o = 2088$ m represents the beams's radius of curvature. The typical speckle patterns obtained for the two OAM states after 2 km atmospheric propagation under these simulation parameters are illustrated in panel I and panel II of figure 10.9. Key second-order statistical features like contrast, correlation,

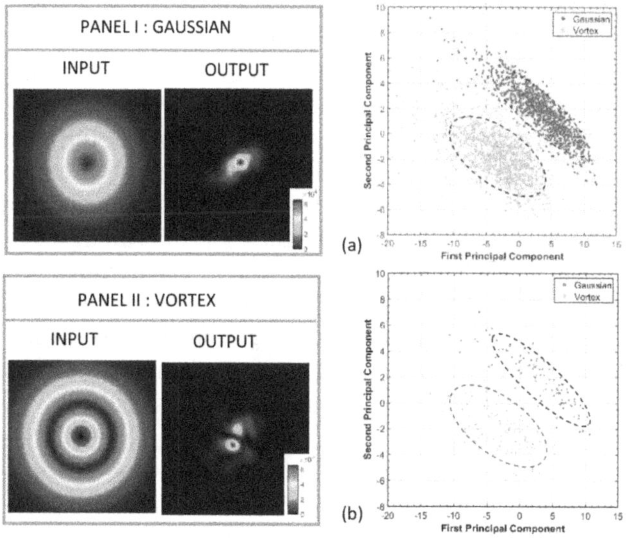

Figure 10.9. Panel I and panel II show typical intensity profiles of focused Gaussian and vortex beams after propagating 2 km through the atmosphere. The input and output planes represent the transmitter and receiver planes, respectively. Each cropped image frame has a width of $1.25 W_o$, (a) shows the 2D PCA plot of the first and second PCs derived from the texture parameters of the training dataset, while (b) illustrates a 2D PCA plot of the first and second PCs obtained from the texture parameters of the test dataset.

homogeneity and energy were extracted using the GLCM methodology and fed as classifiers to train a PCA model.

Figure 10.9(a) illustrates the 2D PCA plots derived from the texture parameters of the instantaneous intensity of vortex and Gaussian beams. In the plots, data points for the vortex beam are marked in green, while those for the Gaussian beam are marked in red. The 2D PCA plot clearly shows that data points corresponding to the same OAM mode form distinct clusters, indicating unique textural characteristics in the speckle patterns of vortex versus Gaussian beams. These clusters are delineated using standard deviational ellipses (SDEs) or covariance error ellipses [21–27]. To further assess the PCA model's performance, a test dataset consisting of 100 separate intensity realizations for both vortex and Gaussian beams is generated under the same propagation conditions. Texture parameters from this test data are used to compute the scores for each principal component (PC) with the trained PCA model, as shown in figure 10.9(b). The plot demonstrates that most of the test data points fall within the SDEs defined by the training dataset. The results demonstrate the effectiveness of the proposed method in distinguishing speckles associated with different OAM values on the PCA plot. By utilizing instantaneous intensity measurements, this method can facilitate efficient real-time identification without relying on adaptive optics.

The above findings highlight the importance of speckle texture diversity for identification of speckles corresponding to various illumination schemes on the PCA plot. It is equally interesting to examine if this method based on speckle texture diversity could be used for characterization of the propagation medium. In this regard, we next discuss the speckle texture diversity-based characterization of atmospheric turbulence for real-time applications.

The atmospheric strength parameter C_n^2 is crucial for describing atmospheric turbulence. This parameter is highly sensitive to temperature fluctuations and changes in altitude, as discussed in chapter 7. The nature of turbulence experienced along slant or vertical propagation paths, as seen in astronomical imaging and horizontal terrestrial propagation, as in FSO communication, is significantly different [28]. Thus, accurately determining the C_n^2 profile along a given atmospheric path is extremely valuable. This knowledge aids in testing directed energy weapon systems for military purposes, assessing and selecting astronomical sites, understanding image degradation over long propagation distances, and evaluating the performance of adaptive optics systems. Therefore, from a system design and testing perspective, understanding real-time variations in C_n^2 is crucial.

Initial researchers have tried to determine C_n^2 values by collecting data on meteorological factors like temperature gradients and wind speed using tools such as resistance thermometers or hot wire anemometers mounted on towers, aircraft, balloons, or tethered blimps [29–32]. Historically, sonic detection and ranging (SODAR) was utilized to study wind profiles [33, 34]. By also measuring air temperature, humidity, and pressure, a carefully calibrated SODAR system can provide C_T^2 profiles. However, these methods require long-term data collection over several months and seasons to accurately estimate C_n^2 profiles for a given location [35].

As these measurements are averaged over time, they can indicate the expected atmospheric turbulence in an area but are not suitable for real-time applications. Additionally, relying solely on meteorological measurements does not fully capture the complex properties of atmospheric turbulence, highlighting the need for newer measurement methods.

Effects such as beam broadening, intensity fluctuations, and angle-of-arrival variations [36] on the beam's intensity and phase profiles can indirectly indicate the strength of atmospheric turbulence and help determine measures such as Fried parameter (r_o) or atmospheric coherence length (ρ_o) used in imaging systems [37]. Various techniques have been developed to use these optical measurements to assess atmospheric turbulence characteristics [38]. Notable methods include optical scintillometers [39–41], Differential image motion monitors [42], Shack–Hartmann wavefront sensors [43–45], LIDAR, SCIDAR [46], path-resolved optical profilers, radiosonde data [47], and systems measuring differential tilt variance [48]. The vertical C_n^2 profile has been studied using techniques like SCIDAR, multi-aperture scintillation sensors [49], SLODAR [50, 51], wavefront outer scale profile monitors [52], and Moon limb profilers [53]. Wave-optics numerical simulations have also been used to analyze atmospheric turbulence by examining backscattered laser light from moving targets [53]. While many of these methods are effective, they often require additional equipment, advanced adaptive optics, large telescopes, and long exposure images. Other approaches, such as multiframe blind deconvolution (MFBD) [54], have been used to process short-exposure intensity images and derive r_o values. Recently, deep learning techniques have been explored to train models to predict turbulence levels from intensity images [55, 56], aiming to improve temporal resolution in turbulence sensing while minimizing the need for extensive data or large hardware setups.

We know that a beam's intensity profile typically experiences more degradation and beam spreading in high turbulence conditions. This leads to higher contrast in the speckle pattern. In comparison, for very low turbulence, the intensity profile at the detector remains largely unchanged. Thus, the speckle pattern generated by a scalar optical beam varies with the level of turbulence encountered during its propagation path. This results in textural variations within the speckle patterns that encode information about the turbulence levels the beam experienced on its way to the detector. The textural properties of speckles, such as contrast, correlation, homogeneity, and energy, can provide novel insights into the level of turbulence or C_n^2 values encountered. Next, we discuss a novel way for characterizing real-time atmospheric turbulence by analyzing the textural properties of speckles generated by a $l = +1$ vortex beam under medium to high turbulence conditions.

Using the random phase screen, speckle for $l = 0$ and $l = 1$ OAM modes (as described in equations (10.13) and (10.14)) are obtained for a 2 km horizontal atmospheric path. Figure 10.10 shows the typical intensity profiles of focused vortex (panel I) and Gaussian (panel II) beams for three different C_n^2 values denoting low (1×10^{-14} m$^{-2/3}$), moderate (5×10^{-14} m$^{-2/3}$) and high (1×10^{-13} m$^{-2/3}$) turbulence levels. It can be seen that as turbulence strength increases, the speckle patterns produced by both beams change noticeably. It is clear that both beams experience

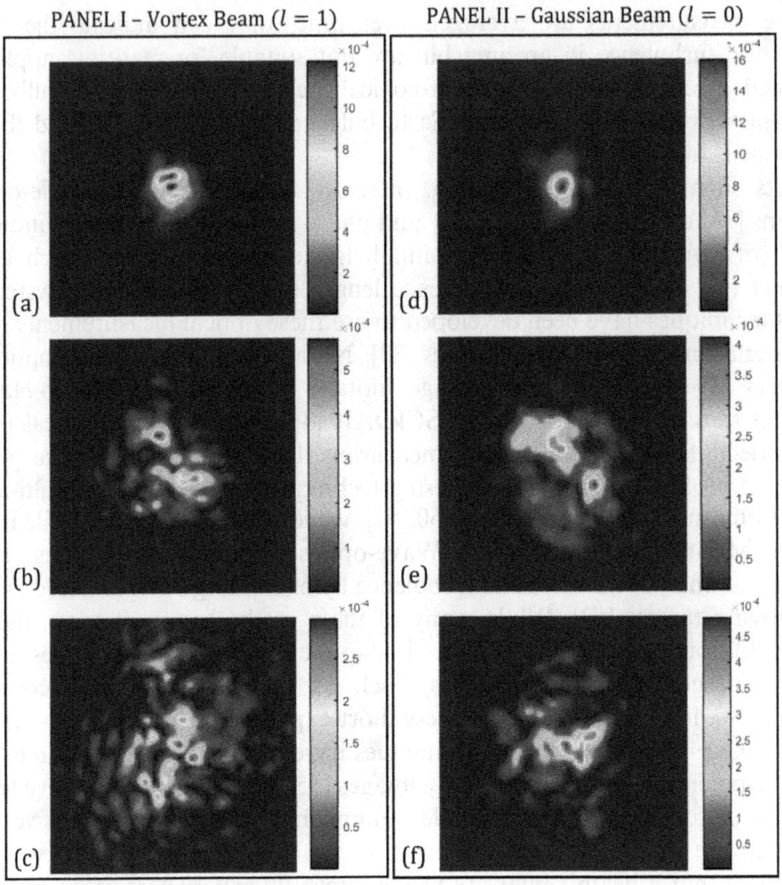

Figure 10.10. The figure displays typical intensity profiles of focused vortex (panel I) and focused Gaussian (panel II) beams after traveling 2 km through different atmospheric turbulence conditions. The atmospheric turbulence strength parameter C_n^2 is set to 1×10^{-14} $m^{-2/3}$ for (a,d), 5×10^{-14} $m^{-2/3}$ for (b,e), and 1×10^{-13} $m^{-2/3}$ for (c,f). Each cropped image frame measures 247×247 pixels, equivalent to a width of $1.75W_0$. Adapted with permission from [57] © 2023 IOP Publishing Ltd.

more intensity degradation and beam broadening with increasing C_n^2 values. This indicates that the spatial distribution of energy in the diffraction speckle is influenced by the level of turbulence encountered during propagation. Therefore, it seems feasible to categorize speckle associated with different C_n^2 values based on the spatial distribution of their intensity variations [57].

In order to train a PCA model, texture parameters are extracted from separate datasets of vortex and Gaussian beams. Each dataset includes 1500 instantaneous intensity images (500 images per turbulence level) collected after the beams had propagated 2 km through specified turbulence levels. Figure 10.11(a) and (b) illustrates the 2D PCA plots obtained for vortex and Gaussian beams. Figure 10.12 shows the 3D PCA plots, which display the first, second, and third PCs for both vortex and Gaussian beams. In both 2D and 3D PCA plots, the data

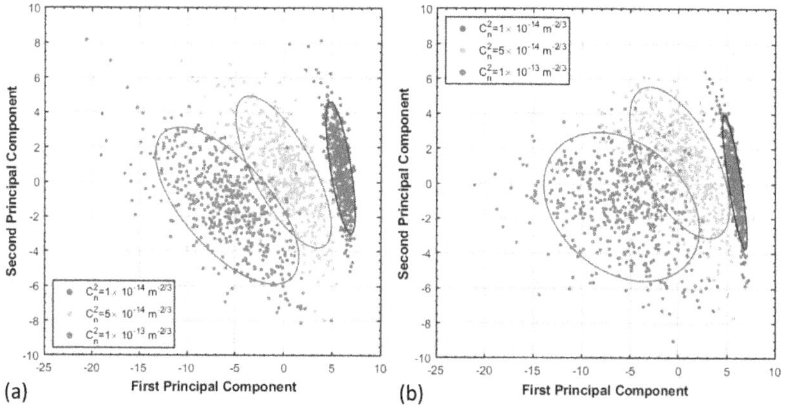

Figure 10.11. The 2D PCA plots present the first and second PCs obtained using the texture parameters of the training dataset for three levels of atmospheric turbulence strength parameter: $C_n^2 = 1 \times 10^{-14}$ m$^{-2/3}$ (blue), $C_n^2 = 5 \times 10^{-14}$ m$^{-2/3}$ (green), and $C_n^2 = 1 \times 10^{-13}$ m$^{-2/3}$ (magenta) for (a) vortex and (b) Gaussian beams. The x and y axes represent the range of values for the first and second PCs. Data points with the same C_n^2 values form clusters, which are outlined by respective SDEs. The clusters for vortex beam show better separation in the PCA plots compared to those for Gaussian beam. Adapted with permission from [57] © 2023 IOP Publishing Ltd.

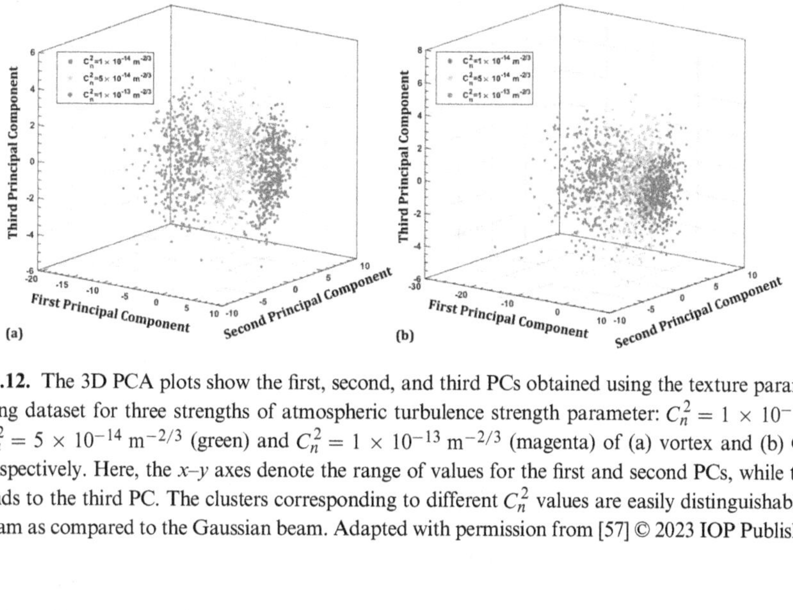

Figure 10.12. The 3D PCA plots show the first, second, and third PCs obtained using the texture parameters of the training dataset for three strengths of atmospheric turbulence strength parameter: $C_n^2 = 1 \times 10^{-14}$ m$^{-2/3}$ (blue), $C_n^2 = 5 \times 10^{-14}$ m$^{-2/3}$ (green) and $C_n^2 = 1 \times 10^{-13}$ m$^{-2/3}$ (magenta) of (a) vortex and (b) Gaussian beams, respectively. Here, the x–y axes denote the range of values for the first and second PCs, while the z-axis corresponds to the third PC. The clusters corresponding to different C_n^2 values are easily distinguishable for the vortex beam as compared to the Gaussian beam. Adapted with permission from [57] © 2023 IOP Publishing Ltd.

points for different values of the refractive index structure function, specifically $C_n^2 = 1 \times 10^{-14}$ m$^{-2/3}$, $C_n^2 = 5 \times 10^{-14}$ m$^{-2/3}$, and $C_n^2 = 1 \times 10^{-13}$ m$^{-2/3}$ are represented in the plots by blue, green, and magenta circles, respectively. The 2D and 3D PCA plots clearly demonstrate that data points with the same C_n^2 values tend to cluster together. These clusters in the PCA plots highlight the distinct textural properties present in the diffraction speckle of beam corresponding to different turbulence strengths. In figure 10.11, the clusters for different C_n^2 values have been identified using SDEs, drawn for a confidence interval of 80%. SDEs are valuable graphical tools for visualizing the spread of 2D datasets, especially when the data have non-zero covariance and are arbitrarily correlated. Figure 10.11 further reveals that data points associated with lower turbulence levels, such as $C_n^2 = 1 \times 10^{-14}\ \mathrm{m}^{-2/3}$, form denser and more compact clusters with a smaller area on the PCA plot. In contrast, data points corresponding to higher turbulence levels, such as $C_n^2 = 1 \times 10^{-13}\ \mathrm{m}^{-2/3}$, create larger and more dispersed clusters with a greater surface area on the PCA plot. Thus, the extent of cluster spread appears to be related to the strength of the atmospheric turbulence.

Another key observation can be made by comparing sub-plots (a) and (b) in figures 10.11 and 10.12 is that the clusters for different C_n^2 values are more distinctly separated on the PCA plot for the vortex beam as compared to the Gaussian beam. For the Gaussian beam, there is considerable overlap in the PCA plot between data points for $C_n^2 = 5 \times 10^{-14}\ \mathrm{m}^{-2/3}$ and $C_n^2 = 1 \times 10^{-13}\ \mathrm{m}^{-2/3}$. The phase singularity inherent in the vortex beam results in speckles with increased spatial frequency information, which aids in more effectively distinguishing between the C_n^2 values on the PCA plot. Thus, the non-zero orbital angular momentum of the vortex beam likely contributes to the distinct separation of data points in the PCA analysis.

We explored how the textural characteristics of speckles generated by propagating $l = +1$ vortex beam can be employed to assess atmospheric turbulence levels. Unlike traditional methods that rely on ensemble-averaged quantities such as scintillation, this technique uses instantaneous intensity measurements, enabling efficient real-time turbulence estimation without additional hardware or complex sensors. By focusing solely on the texture properties of the degraded beam's intensity profile, this method can provide swift real-time estimates of C_n^2 values without the need for adaptive optics correction. Thus, the speckle-texture diversity-based approach can offer a promising new way to monitor atmospheric turbulence in real time, bypassing the need for complex instrumentation and extensive data processing. The speckle-texture-based classification was described here using the PCA technique for ease of understanding and interpretability. We believe that the methodology can be implemented with advanced deep learning networks as well.

References

[1] Goodman J W 1975 *Statistical Properties of Laser Speckle Patterns* ed J Dainty (Berlin: Springer)

[2] Goodman J W 2015 *Statistical Optics* 2nd edn (New York: Wiley)

[3] George N and Jain A 1972 Speckle reduction using multiple tones of illumination *Appl. Opt.* **12** 1202–12

[4] Roggemann M C and Welsh B M 1996 *Imaging Through Turbulence* (CRC Press)

[5] George N and Jain A 1974 Space and wavelength dependence of speckle intensity *Appl. Phys.* **4** 201–12

[6] Mandel L and Wolf E 1995 *Optical Coherence and Quantum Optics* (Cambridge University Press)

[7] Wang W, Hanson S G and Takeda M 2022 Autocorrelation functions and power spectral densities of the stokes parameters in a polarization speckle pattern *JOSA* A **40** 165–74
[8] Kumar B, Lochab P, Kayal E B, Ghai D P, Senthilkumaran P and Khare K 2022 Speckle in polarization structured light *J. Mod. Opt.* **69** 47–54
[9] Hung C C, Song E and Lan Y 2019 *Image Texture Analysis* (Cham: Springer)
[10] Haralick R M and Shapiro L G 1992 *Computer and Robot Vision* (Reading, MA: Addison-Wesley)
[11] Haralick R M, Shanmugam K and Dinstein I 1973 Textural features for image classification *IEEE Trans. Syst. Man Cybern.* **3** 610–21
[12] Haralick R M 1979 Statistical and structural approaches to texture *Proc. IEEE* **67** 786–804
[13] Levine M D and Nazif A M 1985 Dynamic measurement of computer generated image segmentations *IEEE Trans. Pattern Anal. Mach. Intell.* **7** 155–64
[14] Materka A and Strzelecki M 1998 Texture analysis methods—a review *COST B11 report* Technical University of Lodz, Institute of Electronics, Brussels
[15] Bharati M H, Jay Liu J and MacGregor J F 2004 Image texture analysis: methods and comparisons *Chemometr. Intell. Lab. Syst.* **72** 57–71
[16] Chamundeeswari V V, Singh D and Singh K 2009 An analysis of texture measures in PCA-based unsupervised classification of sar images *IEEE Geosci. Remote Sens. Lett.* **6** 214–8
[17] Zhu C and Yang X 1998 Study of remote sensing image texture analysis and classification using wavelet *Int. J. Remote Sens.* **19** 3197–203
[18] MathWorks 2024 Texture analysis using the gray-level co-occurrence matrix (glcm) https://in.mathworks.com/help/images/texture-analysis-using-the-gray-level-co-occurrence-matrix--glcm.html
[19] MathWorks 2024 graycomatrix https://in.mathworks.com/help/images/ref/graycomatrix.html
[20] MathWorks 2024 kmeans https://in.mathworks.com/help/stats/kmeans.html
[21] Raine J W 1978 Summarizing point patterns with the standard deviational ellipse *JSTOR* **10** 32–3
[22] Gong J 2002 Clarifying the standard deviational ellipse *Geograph. Anal.* **34** 155–67
[23] Welty Lefever D 1926 Measuring geographic concentration by means of the standard deviational ellipse *Am. J. Sociol.* **32** 88–94
[24] Yuill R S 1971 The standard deviational ellipse; an updated tool for spatial description *Geografiska Annaler: Series B Hum. Geogr.* **53** 28–39
[25] Härdle W K and Simar L 2019 *Applied Multivariate Statistical Analysis* 5th edn (Cham: Springer)
[26] Furfey P H 1927 A note on Lefever's "standard deviational ellipse" *Am. J. Sociol.* **33** 94–8
[27] Wang B, Miao Z and Shi W 2015 Confidence analysis of standard deviational ellipse and its extension into higher dimensional Euclidean space *PLoS ONE* **10** e0118537
[28] Strohbehn J W 1968 Line-of-sight wave propagation through the turbulent atmosphere *Proc. IEEE* **56** 1301–18
[29] Bufton J L, Minott P O, Fitzmaurice M W and Titterton P J 1972 Measurements of turbulence profiles in the troposphere *J. Opt. Soc. Am.* **62** 1068–70
[30] Barletti R, Ceppatelli G, Paternò L, Righini A and Speroni N 1976 Mean vertical profile of atmospheric turbulence relevant for astronomical seeing *J. Opt. Soc. Am.* **66** 1380–3
[31] Barletti R, Ceppatelli G, Paterno L, Righini A and Speroni N 1977 Astronomical site testing with balloon borne radiosondes–results about atmospheric turbulence, solar seeing and stellar scintillation *Astron. Astrophys.* **54** 649–59

[32] Azouit M and Vernin J 2005 Optical turbulence profiling with balloons relevant to astronomy and atmospheric physics *Astron. Soc. Pac.* **117** 536

[33] Chan P W 2008 Measurement of turbulence intensity profile by a mini-sodar *Meteorol. Appl.* **15** 249–58

[34] Vogt S and Thomas P 1995 Sodar—a useful remote sounder to measure wind and turbulence (Third Asian-Pacific Symposium on Wind Engineering) *J. Wind Eng. Ind. Aerodyn.* **54–55** 163–72

[35] Peralta A, Nelson C and Brownell C 2023 Seasonal changes in atmospheric optical turbulence in a near maritime environment using turbulence flux measurements *Atmosphere* **14** 73

[36] Roggemann M C and Welsh B M 1996 *Imaging Through Turbulence* (Boca Raton, FL: CRC Press)

[37] Li M, Zhang P and Han J 2022 Methods of atmospheric coherence length measurement *Appl. Sci.* **12** 2980

[38] Vorontsov M A, Lachinova S L and Majumdar A K 2016 Target-in-the-loop remote sensing of laser beam and atmospheric turbulence characteristics *Appl. Opt.* **55** 5172–9

[39] Frehlich R G and Ochs G R 1990 Effects of saturation on the optical scintillometer *Appl. Opt.* **29** 548–53

[40] Tunick A 2007 Statistical analysis of measured free-space laser signal intensity over a 2.33 km optical path *Opt. Express* **15** 14115–22

[41] Polnau E and Vorontsov M A 2021 Atmospheric turbulence characterization using a neuromorphic camera-based imaging sensor *J. Opt.* **23** 125608

[42] Hanna R, Brown D M, Brown A and Baldwin K 2022 Measuring atmospheric turbulence along folded paths using a laser-illuminated differential image motion monitor *Appl. Opt.* **61** 9646–53

[43] Silbaugh E E, Welsh B M and Roggemann M C 1996 Characterization of atmospheric turbulence phase statistics using wave-front slope measurements *J. Opt. Soc. Am.* A **13** 2453–60

[44] Sauvage C, Robert C, Mugnier L M, Conan J M, Cohard J M, Nguyen K L, Irvine M and Lagouarde J P 2021 Near ground horizontal high resolution cn2 profiling from Shack-Hartmann slopeand scintillation data *Appl. Opt.* **60** 10499–519

[45] Andrade P P, Garcia P J V, Correia C M, Kolb J and Carvalho M I 2018 Estimation of atmospheric turbulence parameters from Shack–Hartmann wavefront sensor measurements *Mon. Not. R. Astron. Soc.* **483** 1192–201

[46] Coburn D, Garnier D and Dainty J C 2005 A single star SCIDAR system for profiling atmospheric turbulence *Optics in Atmospheric Propagation and Adaptive Systems VIII* **vol 5981** ed K Stein (Bellingham, WA: International Society for Optics and Photonics, SPIE) 59810D

[47] Ko H C, Chun H Y, Wilson R and Geller M A 2019 Characteristics of atmospheric turbulence retrieved from high vertical-resolution radiosonde data in the United States *J. Geophys. Res. Atmos.* **124** 7553–79

[48] Whiteley M R, Washburn D C and Wright L A 2002 Differential-tilt technique for saturation-resistant profiling of atmospheric turbulence *Adaptive Optics Systems and Technology II* **vol 4494** ed R K Tyson, D Bonaccini and M C Roggemann (Bellingham, WA: International Society for Optics and Photonics, SPIE) pp 221–32

[49] Els S G *et al* 2008 Study on the precision of the multiaperture scintillation sensor turbulence profiler (mass) employed in the site testing campaign for the thirty meter telescope *Appl. Opt.* **47** 2610–8

[50] Butterley T, Wilson R W and Sarazin M 2006 Determination of the profile of atmospheric optical turbulence strength from SLODAR data *Mon. Not. R. Astron. Soc.* **369** 835–45 05

[51] Goodwin M, Jenkins C and Lambert A 2007 Improved detection of atmospheric turbulence with slodar *Opt. Express* **15** 14844–60

[52] Ziad A, Maire J, Borgnino J, Dali A, Berdja A, Ben A, Martin F and Sarazin M 2010 MOSP: monitor of outer scale profile *1st AO4ELT Conf. - Adaptive Optics for Extremely Large Telescopes* 03008

[53] Kulikov V A, Lachinova S L, Vorontsov M A and Gudimetla V S R 2020 Characterization of localized atmospheric turbulence layer using laser light backscattered off moving target *Appl. Sci.* **10** 6887

[54] Webb A J, Roggemann M C and Whiteley M R 2021 Atmospheric turbulence characterization through multiframe blind deconvolution *Appl. Opt.* **60** 5031–6

[55] Vorontsov A M, Vorontsov M A, Filimonov G A and Polnau E 2020 Atmospheric turbulence study with deep machine learning of intensity scintillation patterns *Appl. Sci.* **10** 8136

[56] Polnau E, Hettiarachchi D L N and Vorontsov M A 2022 Electro-optical sensors for atmospheric turbulence strength characterization with embedded edge AI processing of scintillation patterns *Photonics* **9** 789

[57] Lochab P, Kumar B, Ghai D P, Senthilkumaran P and Khare K 2024 Real time characterization of atmospheric turbulence using speckle texture *J. Opt.* **26** 015602

IOP Publishing

Orbital Angular Momentum States of Light (Second Edition)
Propagation through atmospheric turbulence
Kedar Khare, Priyanka Lochab and Paramasivam Senthilkumaran

Appendix A

Annotated computer code for beam propagation through turbulence

In this appendix we provide annotated code with brief additional explanations for simulation of arbitrary beam profiles through atmospheric turbulence by the split-step method discussed in chapter 8. The codes used here have been employed for generating a number of beam profile diagrams in this book. For beginning researchers as well as experts, the codes provided here can serve as a handy tool for experimentation and for trying out new ideas after their suitable modification. The codes provided here work in the MATLAB programming environment as well as its Open Source clone GNU Octave. The high level programming languages used by both MATLAB and Octave are straightforward to follow and the reader is expected to know the basics of these programming tools that are now often used for teaching.

A.1 Initialization of parameters

We first need to initialize all the variables for turbulence parameters, the 2D sampling grid, wavelength, beam waist to be used, etc. These are defined below and are self-explanatory. The units used for each numerical quantity are specified in the comments that are shown after % sign in each of the code lines.

A.1.1 Turbulence parameters (scales and strength)

```
Lo = 3;             % Outer turbulence scale [m]
lo = 0.01;          % Inner turbulence scale [m]
Cn2 = 1*10^(-14);   % Cn2 value [m^(-2/3)]
```

doi:10.1088/978-0-7503-5959-7ch11

A.1.2 Sampling grid parameters

```
delx = 0.0024;      % 2D sampling interval for computational grid [m].
% The sampling interval must be less that half of the inner scale (lo/2).

N = 512;            % Number of sample points along x and y directions

D = N*delx;         % Physical length of computational window [m].

% Definition of (x,y) and spatial frequency (fx,fy) grids based on the
% sampling interval.

[x,y] = meshgrid(-D/2+delx/2:delx:D/2-delx/2);
[fx,fy] = meshgrid( -1/2/delx :  1/N/delx :  1/2/delx - 1/N/delx );

% Polar coordinates
 r = sqrt(x.^2+y.^2);
theta = atan2(y,x);
rho = sqrt(fx.^2 + fy.^2);
thetaf = atan2(fy,fx);
```

A.1.3 Optical beam parameters

The parameters like wavelength, beam waist and propagation distance are defined next.

```
lambda = 1.5e-6;            % Wavelength [m]
k = 2*pi/lambda;            % Wave number [m^(-1)]
Wg = 0.1;                   % Gaussian beam waist [m]
zo = (pi*Wg^2)/lambda;      % Rayleigh range
L = 2e3;                    % Total propagation distance [m]
```

We need to estimate the total beam spread on propagation through turbulence of given strength over distance L. It must be ensured that the beam stays well within the computational window. Otherwise the sampling grid parameters must change or the distance over which the propagation can be performed needs to be reduced. This is done via beam spread estimated using relations based on the Fried parameter.

```
rfried=(0.42*k^2*Cn2*L)^(-3/5);
n = (N*k*rfried^2)/(4L);
```

As per equation (8.23), we need to make sure that the quantity (n) above is greater than 4. As a second check, one may estimate the beam width after propagation by distance `L` through turbulence by finding out the beam waist as per equaton (8.29):

```
W = sqrt(Wg∧2 * (1 + (L/zo)∧2) + (2*L/k/rfried)∧2);
```

We can place a check here if the estimated beam spread `2W` will be well within the physical computational window size equal to `N*delx`. Generally if the beam has an initial concave curvature the beam will reduce in size till the focus distance and beam spreading beyond the computational window should not be an issue till the focusing distance is reached. If the above conditions are not valid, the sampling grid parameters or the distance `L` need modification. A Gaussian beam to be propagated through turbulence may be specified as:

```
u = exp(- r.∧2/W∧2–i*pi*r.∧2/(lambda*F));
```

Here an additional concave curvature with radius of curvature `F` has been added on the beam in order to focus it. It has already been noted in equaton (9.12) that in free space, a concave curvature with radius of F leads to focusing of beam closer to the input plane. A focused vortex beam in the OAM state l may be defined as:

```
uoam = A*r.∧l * exp(- r.∧2/W∧2–i*pi*r.∧2/(lambda*F) +
i*l*theta);
```

The fields `u` (or `uoam`) above may be normalized simply by using the Frobenius norm function, so that, they now have unit energy:

```
u = u/norm(abs(u), 'fro');
```

Other forms of initial fields (e.g. Bessel beams, flat-top beams, etc) may be defined in straightforward manner by representing them over the (x,y) grid defined above. The field u will in general be a complex valued matrix of size `N × N`.

A.1.4 Number of phase screens

The number of phase screens that are required for simulation is the next important part of the code. As discussed in section 8.2.3, point 5, we note that we need to place the screens periodically over a distance `delz`, such that, for this distance the Rytov variance has a numerical value of less than 0.1 and also that the Rytov variance over the distance `delz` is less than 10% of the total Rytov variance over distance `L`. This can be achieved by slowly reducing the propagation distance `L` till both the conditions are satisfied. In the code below, the distance `L` is reduced in steps of `deltaL = 10` m. Finally the number of screens are estimated as rounded up integer value of `(L/L1)` where `L1` is the distance which satisfies both the required conditions.

```
totalrytov = 1.23*Cn2*k∧(7/6)*L∧(11/6);
flag1 = 0;
flag2 = 0;
L1 = L;
deltaL = 10;

while(flag1 == 0 || flag2 == 0)

L1 = L - deltaL;
rytovL1 = 1.23*Cn2*k∧(7/6)*L1∧(11/6);

if(rytovL1 < 0.1)
flag1 = 1;
end
if(rytovL1 < 0.1*totalrytov)
flag2 = 1;
end

end % End of while loop

numscreens = ceil(L/L1);
delz = L/numsreens;    % Distance between two random phase screens.
```

The distance `L1` has been chosen as 10 m in the above code. It may be changed to a larger value in order to speed up this step.

A.1.5 Generation of phase screen using FFT method

Having decided on the sampling grid and the number of phase screens needed based on the turbulence strength, we now need to generate the random phase screens that are required for the split-step method. The fast Fourier transform (FFT) method for phase screen generation has been explained in section 8.3.2. This method requires us to specify the phase spectrum and in the code provided below we use two choices, viz., Kolmogorov and von Kármán. Their definitions are as follows:

```
fo = 1/Lo;
fm = 5.92/(2*pi*lo);
A0 = 2*pi*k∧2 * delz;
% Kolmogorov Spectrum
spectrum1 = 0.033*Cn2*(rho).∧(-11/3)*(2*pi)∧(-11/3);
% von Karman Spectrum
spectrum2 = ...
0.033*Cn2*(f.∧2+fo∧2).∧(-11/6).*(2*pi)∧(-11/3).*exp(-f/fm).∧2);
```

The power spectral density for the FFT method is now defined as:

```
PSD = A0*spectrum;           % Here spectrum1 or spectrum2 may be used.
PSD(N/2+1, N/2+1) = 0;     % Turning dc component to zero.
```

Now we generate the actual phase screen as follows:

```
  Phi   =   2*pi/(N*delx)    *    (randn(N,N)    +    1i*randn(N,N)).*
(PSD).∧(0.5);
  phi = real(fftshift(ifft2(ifftshift(Phi)))).*N.*N;
  phasescreen = exp(i*phi);
```

The phase function generated with FFT method still needs the sub-harmonic component which is generated as explained next.

A.1.6 Generation of sub-harmonic phase screen

We will generate phase screens up to a level of `p` as explained in section 8.3.3. The main idea is to upsample the zero frequency pixel and evaluate the sub-harmonic phase screen `phisub` as a local discrete Fourier transform over the upsampled frequency locations. The self-explanatory code is very similar to the code for FFT based screen generation.

```
  phisub = zeros(N,N);
  delf = 1/(N*delx);

  for p = 1:3
  delfnew = delf/(3)∧p;
  [fxnew, fynew] = meshgrid(-1:1);
  fxnew = fxnew*delfnew;
  fynew = fynew*delfnew;
  rhonew = sqrt(fxnew.∧2 + fynew.∧2);
  A0 = 2*pi*k∧2 *delz;
  spectrum2 = ...
  0.033*Cn2*(fnew.∧2+fo∧2).∧(-11/6)*(2*pi)∧(-11/3).*exp
(-(rhonew/fm).∧2);
  PSDsub = A0*spectrum2;
  PSDsub(2,2) = 0;
  Phisub  =  2*pi*delfnew*(randn(3,3)  +  1i*randn(3,3)).*sqrt
(PSDsub);
  phisub1 = zeros(N,N);

  for jj = 1:9
  phisub1  =  phisub1  +  Phisub(jj)*exp(i*2*pi*(fxnew(j)*x  +
fynew(j)*y));
```

```
end

phisub = phisub + phisub1;

end

phisub = real(phisub) - mean(real(phisub(:)));
```

The phase screen generated using the FFT-based method and the sub-harmonic part are added to get the total phase screen at any given location along the propagation path. So we have:

```
phasescreen = exp(i*phi + i*phisub);
```

In cases when the propagation path is slanted or vertical, the numerical value of C_n^2 as the beam propagates will be variable and this aspect may be readily incorporated in the computation by using the appropriate C_n^2 in the definition of both the FFT-based and sub-harmonic phase screens.

A.1.7 Free-space propagation between two random phase screens

The free-space propagation of scalar field between two random phase screens may be evaluated using the angular spectrum method, as explained below. We will denote by `uinit` the field arriving at a phase screen and by `u` the field after transmission through the phase screen.

```
u = uinit.*phasescreen;
U = fftshift(fft2(ifftshift(u)));
alpha = sqrt(k∧2-4*pi∧2*(fx.∧2 + fy.∧2));
H = exp(i*delz*alpha);
fmax = 1/lambda * 1/sqrt(1 + (2*delz/N/p)∧2);
LPfilter = (fx.∧2 + fy.∧2 <= fmax∧2);
U = U.*H.*LPfilter;
udelz = fftshift(ifft2(ifftshift(U)));
```

The field `udelz` acts as `uinit` for the next segment in the split-step method.

A.1.8 Simulation of propagation of vector beams through turbulence

A vector beam may be thought of as a superposition of two scalar beam functions in orthogonal polarizations. For example, two different scalar beams `u1` and `u2` may be propagated over long paths through turbulence by the split-step method defined above. Since the cross-talk between the polarizations is nominal in long-range propagation through turbulence, the total irradiance profile of a vector beam is obtained by summing the irradiances for the two scalar parts.

```
Itotal = abs(u1final).∧2 + abs(u2final).∧2;
```

Here `u1final` and `u2final` are field profiles for the two scalar components obtained after propagation simulation using the split-step method. If the two scalar beams are derived from the same initial laser source, the state of polarization of the vector beam may be of interest. The state of polarization may be determined pixel-by-pixel using the computed numerical values of `u1final` and `u2final`. Finally the polarization basis used may be the *x*–*y* or Cartesian basis or the right circular polarization–left circular polarization basis (or any other orthogonal basis states) and the conversion between the two basis states may be performed by appropriate linear combinations of pixels of `u1final` and `u2final`. Note that the total irradiance of the vector beam as computed above is independent of the choice of polarization basis.

www.ingramcontent.com/pod-product-compliance
Lightning Source LLC
Chambersburg PA
CBHW080538220326
41599CB00032B/6301